T0186811

Natural Stone and World Heritage: Salamanca (Spain)

Natural Stone and World Heritage

Edited by Dolores Pereira
Department of Geology,
University of Salamanca, Salamanca, Spain

ISSN : 2640-0162
eISSN: 2640-0170

Volume 1

Natural Stone and World Heritage: Salamanca (Spain)

Dolores Pereira

*Department of Geology, University of
Salamanca, Salamanca, Spain*

CRC Press
Taylor & Francis Group
Boca Raton   London   New York   Leiden

CRC Press is an imprint of the
Taylor & Francis Group, an **informa** business

A BALKEMA BOOK

Cover: The Frog on the skull.

The frog is one of the most popular symbols of Salamanca. It is a small carving (only a few centimetres long) to be found in the intricate and elaborate "Rich Façade" of the University. Several legends are associated with the Frog, but the most popular is related to the good luck it is supposed to give new students coming into the city if they are able to spot it with no help.

Photograph by Vicente Sierra Puparelli.

CRC Press/Balkema is an imprint of the Taylor & Francis Group, an informa business

© 2019 Taylor & Francis Group, London, UK

Typeset by Apex CoVantage, LLC

Library of Congress Cataloging-in-Publication Data
Applied for

Published by: CRC Press/Balkema
    Schipholweg 107C, 2316 XC Leiden, The Netherlands
    e-mail: Pub.NL@taylorandfrancis.com
    www.crcpress.com – www.taylorandfrancis.com

ISBN: 978-1-138-49954-6 (hardback)
ISBN: 978-1-351-01335-2 (eBook)

To the memory of my mother.
To the memory of my father.

Contents

Acknowledgements

The continuing support of IUGS to the Heritage Stones Subcommission, and UNESCO, through the IGCP program and the IGCP-637 project (Designation of Global Heritage Stones), has made possible to advance the study and diffusion of information on the heritage stones of Salamanca, as well as many other places around the world.

Brian Marker is greatly acknowledged for his critical review of the first draft and his help in putting the English text into shape.

Vicente Sierra Puparelli is acknowledged for his help in photographing the heritage stones in Salamanca to the most artistic level.

Chapter 1

Introduction

The term "natural stone" includes all those "stony" products traditionally used by people in the construction industry, including those used in decoration, sculpture and outdoor and indoor flooring and walls. The applications of these noble materials are diverse. New products and applications are continuously being found, but the original stone used in the construction of our architectonic heritage is undoubtedly the best to be used in the repair and maintenance of that heritage.

Natural stone has been used in construction for thousands of years. Architects throughout time have selected stone both in terms of aesthetics and also for its durability. Magnificent constructions still remain from Roman times and many other ancient cultures, and natural stone is one of the main components of structures of many World Heritage Sites and cities.

Ancient cultures, if possible, sourced stones from the surroundings of the construction site. However, some of their most important monuments used stones from many other places, taking advantage of cheap·manual labour, sometimes including slavery. Such was the case with the Egyptian "Porfido Rosso Antico", a red porphyry greatly appreciated by the Romans. The Empress Cleopatra decided to change the symbolic blue colour to red, and this fashion was followed by other leaders, from Emperors to Popes. Even Napoleon liked to be surrounded by the red porphyry (Wikström *et al.*, 2015). However, although the tombstone of the French emperor is widely thought of being made of red porphyry, in fact it is made of Russian red quartzite, together with Carrara marble, serpentinite, black marble, red marble and other stones. Touret and Bulakh (2016) described the complicated completion of Napoleon's tomb that, ironically, ended up with a stone from the country of one of his major opponents. The Egyptian red porphyry extraction was initially concentrated in Egyptian quarries that had been active from AD 20–400, but, during the Middle Ages, the importing of porphyry to Europe ceased, so alternatives had to be found. Not only the Russian red quartzite found a major place in artistic work. Other red stones, like the Dala (Älvdalen) porphyries from Sweden, started to be appreciated and well known in a society that had finally banned slavery (Wikström *et al.*, 2015).

Many different cultures used stone to build constructions that today are recognised as World Heritage Sites. For example, the Taj Mahal in India (adopted in 1983), built with Makran marble; the Alhambra Palace in Spain (adopted in 1984), built with green and white marble from Macael (Fig. 1.1); the city of Evora in Portugal (adopted in 1986), where splendid buildings and sculptures are made of Estremoz marble; and Salamanca, the Golden City in Spain (adopted in 1988), with monuments of Romanesque, Gothic, Moorish, Renaissance and Baroque style made of Villamayor sandstone. All those stones are recognised today for their important heritage value, by being designated, or candidates for being designated, as Global Heritage Stone Resources (GHSRs).

GHSR designation can be achieved after following a strict evaluation procedure (Marker, 2015). The designation also aims to develop internationally accepted standard approaches to the reporting of technical and aesthetic characteristics of natural stones used for the repair and maintenance of historic buildings, monuments and structures as well as for new construction. Formalisation will help to increase awareness of the potential uses of various GHSRs and provide important information for those engaged in using stone for repair and maintenance. Stones that have been used in important heritage construction and sculptural masterpieces, as well as in utilitarian (yet culturally important) applications, are obvious candidates for GHSR status.

This designation can, if properly disseminated, create increased awareness of available and appropriate natural stone amongst professional workers in geology, engineering, architectural and artistic work, in stone/building conservation and among the general public. In addition, the designation can enhance international cooperation for research on, and documentation of, natural stone resources. This has already been demonstrated by numerous enthusiastic contributions to specific sessions in international meetings and publications dedicated to this topic (e.g. Pereira *et al.*, 2015). Success of

Figure 1.1 The Alhambra palace in Granada, built with local marble and gypsum-plaster in the ornamentation of its rooms and façades. The Alhambra, together with the Albayzín, is located in the mediaeval part of Granada. Outside there are the magnificent gardens of the Generalife, the rural residence of the emirs that governed this part of Spain during the 13th and 14th centuries. These three places constitute the World Heritage Site.

the GHSR designations should also help to encourage proper management of natural stone resources including the future protection of important dimension stone resources from sterilisation by other forms of development (Cooper *et al.*, 2013).

The procedure for recognition of a stone as a GHSR is supervised by the Heritage Stones Subcommission (HSS), a working group within the International Union of Geological Sciences (IUGS). It involves several steps after the considered stone candidate meets most of the following requirements:

- Wide-ranging geographic application and/or historic use for a period of at least 50 years, but preferably centuries.
- The potential candidate dimension stone should also have been utilised in significant public or industrial projects.
- There should be wide recognition of the stone for its cultural importance, including association with identity or a significant contribution to architecture at international, national, regional or local level.
- It is beneficial that stone remains available in quarries, even if these quarries are not currently active.

Other potential benefits (including cultural, scientific, architectural, environmental, commercial) are also be considered.

In order to present a stone for candidacy as a GHSR, a checklist must be attached to the proposal, with the following content:

- Formal stone name.
- Stratigraphic (or geological) name.
- Other names (names given to different types or variants of the stone).
- Commercial designations (additional commercial names used to market the stone).
- Area of occurrence (geographic area where the considered stone occurs in nature, including a map with location of outcrops or quarries).
- Geological age and geological setting (details of sedimentary/basin/fold belt, tectonic domain, igneous activity etc. that place the considered stone in a wider geological perspective).
- Petrographic name (technical name of stone as determined by scientific assessment).
- Primary colour(s) and aesthetics of the stone.

- Natural variability.
- Composition (distinguishing mineralogical characteristics).
- Geotechnical properties (physical and mechanical characterisation. At least the most useful data: water absorption, capillarity absorption, compression strength, flexural strength, porosity, etc.).
- Suitability (assessment on utilisation, for example, masonry, dimensional blocks, sculpting stone, roofing, etc.).
- Vulnerability and maintenance of supply (availability of future supply including possibility of constraints on supply).
- Historic use and geographic area of utilisation (historic and geographic utilisation of the nominee especially in significant heritage or archaeological applications).
- Heritage utilisation (an extensive list of the significant buildings, monuments, sculptures, etc., including dates of construction).
- Related heritage issues (information on significant heritage issues that affect the stone, for example, alternative heritage listing of buildings or quarry areas associated with the stone, supporting museums, sculpture parks, etc.).
- Other designations (optional) (proposal of additional designations, for example the epithet "Classic World granite/marble/ etc.", "International Decorative Stone Icon", etc.).
- Related dimension stones (other dimension stones that are related geographically, geologically or utilised together with the proposed stone).
- Principal literature related to the proposed stone (major scientific papers, books and popular literature dealing with the stone).
- Images (images, historic photos and line illustrations for publication).

Also, any other relevant information can be included, such as other benefits that will be achieved through the designation of the stone as GHSR.

This checklist is normally part of a publication in a widely recognised international journal and can serve as the proposal itself. With this step, it becomes possible that many stones from around the world that were described in obscure publications or in minority or not widely accessible languages can reach wider audiences so that specific stones are better known worldwide.

The existence of several natural stones that share geographical location, historical buildings and/or geological history give place to another concept: Global Heritage Stone Province (GHSP) (Pereira and Cooper, 2015). All these natural stones do not have to hold the GHSR designation.

There are many other long-term benefits coming from this designation. For example, for geologists, GHSR designation facilitates the formalisation of the characteristics of natural stone materials for professional purposes in an internationally accepted context. Also, the existence of GHSR in a site, or a quarry well known for its use in the construction of heritage, can help with the designation of an extraction location or locations as a World Heritage Site. Many stone-related sites on the list are considered as potential World Heritage Sites, such as the Marble Basin of Carrara (Fig. 1.2), which is already on the UNESCO tentative list (see https://whc.unesco.org/en/tentativelists/5004/).

Figure 1.2 Carrara Quarries in northern Tuscany (Italy). The quarrying of marble in this area started in the 1st century AD, under Roman rule. Today around 30 quarries from the Roman period can be found in the area. The extraction of the marble is carried out at an altitude between 200 and 1600 m above sea level. The quarrying areas are part of a Regional Park. For this reason, and to respect the environmental laws of the Park, several underground quarrying operations have started (Angotzi *et al.*, 2005). Even so, continuous conflicts among marble companies and environmentalists take place, due to the highly disturbed environment around the quarrying sites (Gentili *et al.*, 2011).

Sometimes the stone-related sites are not the main objective of a world heritage significance but, rather, a larger area. For example, we can take the case of three Egyptian quarry sites: the New Kingdom Unfinished Obelisk in Aswan ("Nubian Monuments from Abu Simbel to Philae"), the Old Kingdom limestone quarries by the Giza pyramids and other pyramid sites ("Memphis and its Necropolis – the Pyramid Fields from Giza to Dahshur") and the New Kingdom limestone quarry close to the Valley of the Kings on the west bank at Luxor ("Ancient Thebes with its Necropolis"). Clearly, the major monuments such as the Obelisk, the pyramids, tombs and temples overshadow the quarries from which the construction materials came, but their value in relation to the creation of the sites should still be recognised. People should be aware that ancient quarries are far more than just a past source of stone but are also very rich archaeological sites in their own right and deserving of study and preservation to provide a more complete perspective on ancient cultures (Harrell and Storemyr, 2009).

The GHSR designation is potentially a mechanism for formally defining a stone type, of a similar manner to the existing European Union legal provisions that protect the local integrity of various foods and drinks from specific areas (Champagne, Cognac, Rioja, Stilton, etc.). If stones are also recognised in that way, then it would protect their reputation by not allowing lower quality and cheaper stones from elsewhere to be marketed under the same names. That is important because, in many cases, these would not meet most of the requirements of the original stone. Designations can also help to increase the awareness of the importance of these important products among professionals, such as architects, who often select the stone only on aesthetic criteria or price rather than the suitability for the specific purpose, as well as the wider public (see Figure 1.3).

Natural stone is a durable construction material, but it is affected by the same processes of weathering by water, wind, frost, heating and biological activity as bedrock exposures of rocks. Inevitable progressive deterioration reflects the nature and properties of the stone, the passage of time and the ambient conditions, both natural and anthropogenic, to which it is exposed. Causes of deterioration of stone are varied, and include cracking and deformation, dissolution, detachment, loss of material through erosion and mechanical damage, effects of discolouration and surface deposits and biological colonisation (e.g. algae, bacteria) (ICOMOS-ISCS, 2008).

Figure 1.3 Understanding the heritage value of stones helps to promote the proper management of natural resources as part of a sustainable world. The Oxford University Museum of Natural History has an interior court surrounded by 30 columns, each made of a different British stone. Each column is labelled with the name of the stone and the place from which it came. This is a good practice to involve a community in the knowledge of the country's resources. This museum also keeps a valuable collection by Faustino Corsi, who built up a collection of over a thousand polished specimens representing the stones that had been used in Ancient Rome.

Source: (Pereira and Marker, 2016a).

Many stone-built structures are in urban areas, so, particularly since large-scale industrialisation, these have been exposed to aggressive attacks by pollutants, accelerating the rate of decay. Because of early industrialisation and large numbers of historically important structures, Western Europe is an instructive area for observing causes of decay and possible approaches to reducing future damage to the historical, cultural and architectural heritage.

Article 4 of the UNESCO "Convention concerning the Protection of the World Cultural and Natural Heritage" states that "Each State Party to this Convention recognises the duty of ensuring the identification, protection, conservation, presentation and transmission to future generations of the cultural and natural heritage" (see whc.unesco.org/en/conventiontext). This can help to protect national heritage and can contribute to income from tourism.

Governments address this obligation in different ways depending on their national priorities and provisions for protecting heritage. However, positive efforts can be undermined by political instability and loss of control. Wars and vandalism have endangered historical areas and sites for centuries, causing damage or, in some cases, complete destruction. Recent and current political instability in some places, including some UNESCO World Heritage Sites, is a matter for continuing concern but can only be solved by conflict resolution. More widely, anthropogenic deterioration, whether conscious or unconscious, or caused by climate and weather (http://whc.unesco.org/en/danger/), must be addressed by good practices for repair and maintenance (Pereira and Marker, 2016b).

Inappropriate maintenance and repair, development projects, inadequate management systems and insufficient legal protection can threaten either important structures or groups of individually less important buildings that, together, constitute significant conservation areas. The rate of deterioration of stone depends on the initial quality and can progress to a condition in which only replacement can secure the future of the building or monument.

Intervention at the right time can preserve, or extend the life of, the cultural heritage, but technically and aesthetically appropriate materials must be selected to retain both visual appearance and the structural integrity of constructions. Inappropriate materials can accelerate future damage. Therefore, the original types of stone should be used for maintenance and repair, but that may be impossible if resources have been exhausted, built over or have otherwise become inaccessible. In that case, detailed and readily accessible technical information is needed to identify the most appropriate alternatives.

Inappropriate actions in the repair and maintenance of buildings still occur widely, even in parts of UNESCO World Heritage Sites and cities. Some examples of problems in these and other locations in Western Europe serve to illustrate the salient issues and are illustrated in this book.

Although contemporary buildings and premises in Salamanca, as in most European cities, use all kinds of natural stone from all over the world (e.g. Larvikite from Norway, Rapakivi granite from

Finland, Charnockite from India, etc.), historic buildings were built only with local stone, but that also varies due to the diversity of geological environments in the province. In this book the different geological materials used in the construction of some of the most important buildings in the city are described and examples of use and misuse of natural stone are also examined.

It needs to be recognised that some quarries from which the stones were extracted are still active. However, others are currently inactive, and experts recommend that these should be preserved as historic quarries so that suitable materials can be made available for particularly important buildings in the city.

It is unavoidable that some technical words must be used, so a glossary is included to explain these terms. Also, a list of further reading is provided, ranging from scientific to historical projects, some technical handbooks relating to UNESCO's activities on World Heritage conservation and examples of stone deterioration and decay that are relevant to repair, maintenance and restoration.

A principal purpose of this book is to stimulate awareness of the importance of stone in the preservation and restoration of a World Heritage Site, such as Salamanca, which achieved the recognition mainly for the homogeneous architecture built with local natural stone, although that has not previously been adequately emphasised.

Chapter 2

World Heritage Sites

Cultural heritage is part of our legacy from the past. Our ancestors built it and took care of it. Now we live with it, and we should pass it on to future generations, as part of a sustainable world. The concept of "world heritage" is of universal application, which makes it so exceptional. World Heritage Sites belong to everybody around the world, independently of where they live.

The United Nations Educational, Scientific and Cultural Organization (UNESCO) seeks to encourage the identification, protection and preservation of cultural and natural heritage around the world considered to be of outstanding value to humanity. This is embodied in an international treaty called the Convention concerning the Protection of the World Cultural and Natural Heritage, adopted by UNESCO in 1972.

UNESCO's world heritage mission is to encourage countries to sign the World Heritage Convention and to ensure the protection of their natural and cultural heritage, encourage states to the convention to nominate sites within their national territory for inclusion on the World Heritage List, encourage states to establish management plans and set up reporting systems on the state of conservation of their World Heritage Sites, help states to safeguard world heritage properties by providing technical assistance and professional training, provide emergency assistance for World Heritage Sites in immediate danger, support public awareness–building activities for world heritage conservation, encourage participation of the local population in the preservation of their cultural and natural heritage and encourage international cooperation in the conservation of our world's cultural and natural heritage (see https://whc. unesco.org/en/about/). Society in general and researchers in particular also have a major role in maintaining the cultural heritage in the best state for future generations and that requires knowledge of the materials that the heritage was built with and the best way of undertaking repair, maintenance or restoration.

At present, UNESCO recognises 1092 World Heritage Sites, including both natural and cultural sites. From these, 845 sites are recognised for their cultural value mainly related to historic city centres, specific historic monuments and sites containing natural stone. Europe is currently the continent with the most World Cultural Heritage Sites (more than 400), followed by Asia and Pacific countries (181), Latin America (96), Arab States (76), Africa (52) and North America (7). In all the regions, there are several cultural sites that are in danger: currently 3 in Europe, 4 in Asia and Pacific countries, 22 in the Arab States, 4 in Africa and 5 in Latin America. This means that if actions do not achieve improvements, the sites can be de-listed from their recognition as World Heritage Sites.

UNESCO publishes a list of World Heritage Sites in danger to inform the international community of conditions that threaten the very characteristics for which a property was included in the World Heritage List, and to encourage corrective actions. A list of World Heritage Sites in danger can be found at https://whc.unesco.org/en/danger/.

The convention considers two cases when a World Cultural Heritage Site is in danger:

- Ascertained Danger: the site is faced with specific and proven imminent danger such as the following:
    - Serious deterioration of materials.
    - Serious deterioration of structure and/or ornamental features.
    - Serious deterioration of architectural or town-planning coherence.
    - Serious deterioration of urban or rural space, or the natural environment.
    - Significant loss of historical authenticity.
    - Important loss of cultural significance.

- Potential Danger: the site is faced with threats which could have deleterious effects on its inherent characteristics. Such threats could be these:
    - Modification of juridical status of the property diminishing the degree of its protection.
    - Lack of conservation policy.

- Threatening effects of regional planning projects.
- Threatening effects of town planning.
- Outbreak or threat of armed conflict.
- Threatening impacts of climatic, geological or other environmental factors.

In fact, some sites have been de-listed recently, and several others seem to be on the same track, due to several of the threats listed. And some of those threats are directly related with the state of the construction stones. Deterioration of the stone derived from anthropogenic activities, whether conscious or unconscious, or caused by climate and weather, can be addressed by good repair and maintenance practices (Pereira and Marker, 2016b). The rate of deterioration of stone depends on the initial quality and can progress to the point when only replacement can secure the future of the building or monument.

Intervention at the right time can preserve, or extend the life of, the cultural heritage, but technically and aesthetically appropriate materials must be selected to retain both visual appearance and structural integrity of constructions. The recommendation is to use the original types of stone for maintenance and repair, but that may be impossible if resources have been worked out, built over or otherwise have become inaccessible. In that case, detailed and readily accessible technical information is needed to identify the most appropriate alternatives.

Inappropriate actions in the repair and maintenance of buildings occur widely, even in parts of UNESCO world heritage cities and sites.

Chapter 3

Salamanca
A World Heritage Site

Salamanca is a city in Spain, capital of the province of the same name, located in the region of Castilla y León, in the Spanish North Plateau. The city has a population of around 150,000 inhabitants, and the province has a population of around 330,000 people.

The historic town of Salamanca was designated as a World Heritage Site by UNESCO in 1988. It is a unique city with many monuments and historical buildings made of local natural stones, from sedimentary to igneous rocks that were (and are) extracted in quarries nearby.

The origins of Salamanca go back 2700 years, coinciding with the Iron Age. The first settlers occupied a hill that overlooked the Tormes river and was surrounded by river channels: San Vicente hill (Macarro and Alario, 2012). This hill was the ideal location for defensive reasons, and the nomad settlers started to build their more permanent dwellings on what today is called "Cerro de San Vicente" (Fig. 3.1).

Those settlers predated the Vettones and Vacceans, pre-Roman cultures. All of these cultures lived around the Duero river, in the western most part of Iberia. There is no documentation of their existence in other parts of the peninsula. But the first settlers did not build their houses with stone, even if they were living just above very good outcrops. They built their houses with mud and adobe. Today the remains of that ancient culture can be found as a protected site. The earliest culture leaving vestiges of the use of stone were the Vacceans: Salamanca province has several zoomorphic sculptures as examples.

But the city's (and province's) most important structures built of stone were due to the Romans: the bridge that crosses the Tormes river and gave access to the city from the south, the wall that

Figure 3.1 San Vicente hill, the site of the first settlement in Salamanca. The hill is made up of opal-cemented conglomerate. Most of the city is built on top of this substratum. On top of the hill, archaeological excavations of that first settlement have been undertaken (Fig. 3.2).

Source: Picture by Vicente Sierra Puparelli.

Figure 3.2 Archaeological excavations at San Vicente hill. Macarro and Alario (2012) describe the complete structure of the settlement and the relation of this community with the development of the city of Salamanca.

Source: Picture by Vicente Sierra Puparelli.

originally surrounded part of the city boundaries and the Roman road that crosses Salamanca connecting it with the north and the south of Iberia (Fig. 3.3). This Roman road is also known as Vía de la Plata (Silver route, although the road has nothing to do with

Figure 3.3 A section of a Roman road in the village of Fuenterroble, about 50 km south of Salamanca. The local stones were used to build this road and the structure is similar to that of today's roads, with base and sub-base courses, made up of different sizes of fragments to improve drainage and reinforce the structural support.

silver), which starts in Seville (Andalucía) and goes all the way north to Santiago de Compostela (Galicia). This is a stone road that was used for ages for trading, including the transport of stone from quarries to the city of Salamanca.

Romans used all the different stones that will be discussed in this book in the construction of the Roman bridge, not only from the sandstones to the granites, but also metamorphic slates appear in its structure.

It was in the Middle Ages that the core of Salamanca as we know it today was established and the first cathedral was erected. During the Modern and Contemporary periods, Salamanca developed into a major city and the second cathedral (the New Cathedral) and many historical buildings were built, with the population of the time moving to what was called "the cathedral hill" (*cerro de la catedral*).

However, some of these magnificent buildings were destroyed by the French army during their occupation in the War of Independence (1807–1812), as they wanted to build new defences, and more damage was caused when the French were driven out of the city. It is documented that what we see today in the historic part of Salamanca is only one fifth of what was built before that war.

Today several buildings and monuments of Salamanca city and province show the degree of destruction that the battles could provoke in what today is called cultural heritage. Thanks to the preservation of those buildings, the adverse material effects of a war remain visible (Fig. 3.4).

The city reached its apotheosis in the 16th century thanks to its prestigious university, which is the oldest in Spain and one of the oldest universities in Europe. The designation as a UNESCO World Heritage Site in 1988 was mainly due to its historical and sustainable architecture, which was built from local stone: opal-cemented conglomerate, sandstone, slate and granite were used from the bottom ashlars to the façades, and were included in some structural elements in different buildings around the city. All of these stones were quarried and brought from nearby areas and they have continued to be used in the restoration of historic buildings, with relative success. All upper façades are made of Villamayor sandstone. Sometimes, the whole building, from the basement to the upper façades, were built of Villamayor sandstone, the use of which indicated the importance of the building's owner in a highly aristocratic society. While sandstone (whether Villamayor or Salamanca sandstone, see description of Sedimentary Stones in Chapter 6) had been used for the original work hundreds of years ago, it has been replaced with granite during repair through the centuries. This is due to the degradation that the sandstone underwent over time. Granite materials generally remain in good condition, although there are some influences that lead to the descaling and fading of the stone, and thus architects should make a rigorous assessment when considering whether to use one granite or other in the restoration of historic buildings, based on scientific studies rather than on budget.

Furthermore, the different granite typologies produce very different materials, so they should not be replaced out of hand. It is always advisable to ask Geology specialists who understand the properties of various rocks.

Figure 3.4 Tower of Ciudad Rodrigo cathedral, 80 km southwest of Salamanca. The impact of cannon balls is evident on the exterior walls of the building.

The widespread use of local stone by the most prominent architects of the time to build Salamanca's historic buildings has contributed to the fact that Villamayor sandstone was designated as Global Heritage Stone Resource (GHSR) by the Heritage Stones Subcommission of the International Union of Geological Sciences (IUGS) after following a strict evaluation protocol. Moreover, one of the granites has been proposed as a candidate for the recognition as a GHSR figure and is in the first steps of the designation procedure. Also, the whole set of Salamanca province lithologies have been proposed as a Global Heritage Stone Province: an area in which several different internationally and nationally important stones occur (Pereira and Cooper, 2013).

The recognition as a World Heritage Site was mainly based on the homogenous construction of the old town using local natural stone and the optimum state of conservation. Buildings in central Salamanca were constructed using Villamayor sandstone and Salamanca sandstone (Nespereira *et al.*, 2010; Pereira and Cooper, 2013) for most of the structures. However, granite was used for lower parts of the buildings after it was realised that the sandstones were not resistant to water absorption and become weak under critically wet conditions (Pereira *et al.*, 2015). In some buildings, a sequence of replacements can still be seen on the lower parts of their façades (Figs. 3.5a and 3.5b).

The Casa de las Conchas is in front of the Clerecía church, originally known as the Royal College of the Company of Jesus; construction of this began during the 17th century in Baroque style using Salamanca sandstone in the lower part and Villamayor sandstone in the upper part of the building. The lower part has deteriorated unevenly, due to the water adsorption through the more porous parts of the stone, and several inappropriate actions took place, including covering the stone with mortar in some parts and replacing the blocks with the wrong stone in the frontage of the church.

Humidity and contamination have had negative influences on the sandstone, leaving some buildings in a very poor state. Mortar was used to disguise the deterioration, a serious mistake because it can react chemically with minerals in the stone, especially where the stone has a high water absorption coefficient. The consequent reactions cause accelerated deterioration as the mortar continues reacting with the sandstone matrix and cement and can lead to

Figure 3.5a On the right-hand side, Casa de las Conchas (the Shells House). It is an urban palace of Gothic and Plateresque styles built between 1493 and 1517, with a mixture of mediaeval, Arabic and Renaissance elements in its interior. Imported Carrara marble was used to build the columns in the upper level. Since 1993 it has been a public library that belongs to the national government. The original construction was in Villamayor sandstone and Salamanca sandstone, but subsequent restorations used granite to replace lower ashlars.

Figure 3.5b Limited understanding of natural stone led the architects in charge of the restoration of the Clerecía church to use various igneous rocks to replace the sandstone. The result is a poor aesthetic effect that could have been easily avoided by awareness of available local material (Pereira and Cooper, 2013).

the complete destruction of the stone. Several specific examples of bad restoration and conservation actions can be seen in Salamanca (Figs. 3.6a and 3.6b).

Doubtful restoration actions occur widely even at many World Heritage Sites or sites that submitted for world heritage status (Pereira and Marker, 2016b). For instance, similar circumstances have been observed in Torino, Italy, where attempts were made to restore the foundations of the main entrance of Palazzo Madama (Fig. 3.7). This palace is part of the Savoy Royal Residences, and was added to the UNESCO World Heritage List in 1997. Similar bad practice has been observed in many buildings in Oxford, United Kingdom, where mortar has also been used to "repair" limestone (Fig. 3.8).

Granite, as well as sedimentary rocks, can be affected adversely by inappropriate coverings, although the result is less dramatic than

Figure 3.6a Monterrey palace, one of the greatest examples of the Italian Renaissance style, was built in the 16th century by Pedro de Ibarra and Pedro de Miguel y Aguirre after the plans of Rodrigo Gil de Hontañón and Fray Martín de Santiago. This represents the great nobility of the Spanish "Golden Age". The building was recognised as a National Historic Monument in 1929. The base levels of the building were constructed with the opal-cemented conglomerate, while the façade used Villamayor sandstone.

Figure 3.6b During the 1960s, lower levels of some buildings were covered with mortar, including the Monterrey palace as well as the Clerecía church. Later restoration actions removed the mortar and substituted the conglomerate with contrasting granite blocks, which should not have been allowed in a historic building of this importance.

Figure 3.7 Palazzo Madama is one of the Savoy Royal Residences, located in Torino. It has been part of the UNESCO World Heritage List since 1997, representing monumental architecture styles between the 17th and 18th centuries, and is an expression of monarchical status. The foundations are made of Foresto marble, which is in fact a fine-grained whitish limestone with many parallel-oriented mica crystals. This characteristic will give way to microcracks and then open fractures due to the preferential absorption of water by these minerals and the consequences after freezing and thawing periods (Marini and Mosseti, 2006; Agostini et al., 2017). The attempt at restoration was done by covering the affected limestone with mortar.

Figure 3.8 Deteriorated limestone in a building on one of the main streets of Oxford; another example of bad "restoration" action.

in the case of sandstone and limestone, at least in the short term. However, it is still aesthetically undesirable (Pereira and Marker, 2016b).

Poor restoration practice can, therefore, involve the poor initial selection of stone, as well as the use of technically or aesthetically incompatible stone during the repair and maintenance of structures. Inappropriate materials are often used because of the lack of awareness of the need for appropriate materials or from securing cheaper materials without appreciating the longer-term consequences of damage and costs or from the inappropriate use of mortar or cement. There is a need to raise awareness and understanding of the importance of good practices for the maintenance and repair of our cultural heritage made of natural stone.

Chapter 4

The buildings in Salamanca

Although impressive buildings were constructed during the 11th and 13th centuries, most historical buildings in Salamanca were built between the 15th and 19th centuries. Most of them are ecclesiastical and aristocratic in nature. A very significant proportion of these buildings are related to the university throughout its 800-year history (since 1218).

In the historic centre of Salamanca, declared a UNESCO World Heritage Site in 1988, one can find historic buildings and monuments with their basement constructed using ashlars of different lithological nature. Their state of conservation depends on the kind of stone used. Some bottom ashlars were built with Salamanca sandstone, which is in fact a conglomerate containing siliceous cement that decreases its water absorption capacity. Other bottom ashlars are built on Villamayor sandstone, as the rest of the building. At other times, the building sits on pieces of granite of different typology and geological origin, almost always from nearby places within the province. When the bottom ashlars are made of sedimentary rocks, whether the conglomerate or the Villamayor sandstone (read about the distinctions between them in Chapter 6), they are severely affected by weathering and anthropic processes. This is mainly as a consequence of the high capillary water absorption coefficient of these types of stone (Fig. 4.1).

Granite bottom ashlars remain in a better state of conservation. That is why granite has almost always been used when restoration works have been carried out. However, the differences among several different granitic materials were not included in the protocols for material restoration tasks, which has resulted in different physical, mechanical and aesthetic mistakes. This is particularly true in the case of a large number of monuments in Salamanca that

Figure 4.1 San Benito church, of Romanic style, was built in the 12th century. The whole building is made of Villamayor sandstone, and, in the lower parts, the effect of capillarity absorption can be seen affecting the stone, with biological colonisation (e.g. lichens). Slates are covering some parts in the attempt to protect the surface of the stone. However, the deterioration is actually due to upward infiltration from underground waters.

used the nodular granite of Martinamor, locally nicknamed *Piedra Pajarilla* ("Little Bird Stone" in English). Prominent architects in past years insisted on the use of this natural stone in their constructions (e.g. Juan de Sagarvinaga, Juan de Alava, Juan Gómez de Mora, Alonso de Covarruvias and others), all of them working on the preservation of the impressive architectonic heritage of Salamanca. These architects insisted on reinforcing the buildings affected by the Lisbon earthquake in 1755 only with the "good Martinamor granite", as in the case of the New Cathedral (Rodríguez de Ceballos, 1978; Portal-Monge, 1988). Granites from Sorihuela and Los Santos were also used, as well as vaugnerite (a type of monzodiorite) from Ledesma, which was erroneously described as *Piedra Pajarilla* in some classical literature (Madoz and Sagasti, 1845), and has been continuously mistaken in more recent papers.

This error probably started the confusion among architects and promoters of restoration when selecting the replacement material (see later for details).

Because Salamanca is surrounded by important granitic massifs of Hercynian age, it is possible to find a diversity of granites in the basements of buildings, which can be difficult to identify, but which were clearly extracted from nearby quarries. There is no extensive body of knowledge about the nature of the different granites and their behaviour, so some of these cannot be distinguished accurately, as they lack distinctive features.

Because Salamanca's historic city centre is a World Heritage Site, it is an essential requirement for the town to preserve the appearance of the historic buildings and monuments (UNESCO, 2016). Therefore, it is very important to respect a protocol in all protection, conservation and restoration actions.

In this book, there is a description of the characteristics of the different lithologies that were identified in the historic architecture of Salamanca, as well as the locations of the quarries where they were extracted. Some technical features of the different rocks are included, together with reference to other publications for further information. These characteristics can be compared to the standard recommended values for the use of the stone for construction. Each stone is accompanied by a photograph of a building in which it is mostly used, although, in the majority of buildings, more than one type of stone has been used. For most stones there is also a photograph of the quarry where the stone was sourced. As a general rule, granite is used as the main construction material for the lower courses (whether in the original construction or after restoration actions), while, in the upper part, Villamayor sandstone is always found. Salamanca sandstone (or opal-cemented conglomerate) is also found in the bottom ashlars of some buildings, and this is the most difficult material to replace because all historical quarries are currently inactive. However, architects are finding a way of reaching a very plausible compromise, as described later.

This book can complement other books and guides about tourism and history in Salamanca, where a complete list of the many historic buildings and monuments can be found.

The ultimate purpose of this book is to trigger a specialised literature of world heritage cities and their construction stones, with the goal of preserving our architectural heritage.

Chapter 5

The stones of Salamanca

It is obvious that the stones we find in the historic buildings of the cities nowadays came preferentially from quarries nearby to avoid transport costs, and those quarries depended on the geology of the surrounding area.

The Iberian plateau is one of the largest outcrops of the European Hercynian chain. This plateau is divided into Palaeozoic and Precambrian series. The rocks that are described in this book, and that served for the construction of the historic buildings of Salamanca, belong to the so-called Central-Iberian zone, with Tertiary sediments and granitic rocks (Díez Balda, 1986).

Salamanca was built on Tertiary materials related to the Duero river. Sedimentary rocks are the most common geological material around the city of Salamanca. Tertiary sandstones from the Palaeocene and Eocene time periods were the preferred initial construction materials for the historic buildings of the city (Nespereira *et al.*, 2010).

Nearby bedrock includes slates formed from clays by increased pressure and temperature during ancient tectonic events. Also, there are Hercynian granites, mostly produced by the partial melting of metasedimentary rocks of Precambrian age (Díez Balda, 1986).

This geology (Figs. 5.1a and 5.1b) allowed for the development of the quarries of sedimentary, metamorphic and igneous rocks that provided the construction materials for the historic buildings and structures of Salamanca city and the villages nearby.

Numerous quarries were established in the local province as the characteristics (or better the behaviour) of the different rocks were discovered. All these quarries are now part of Salamanca's landscape and should be preserved as witnesses of important traditions and local activities: these are also part of the cultural heritage (Douet, 2015).

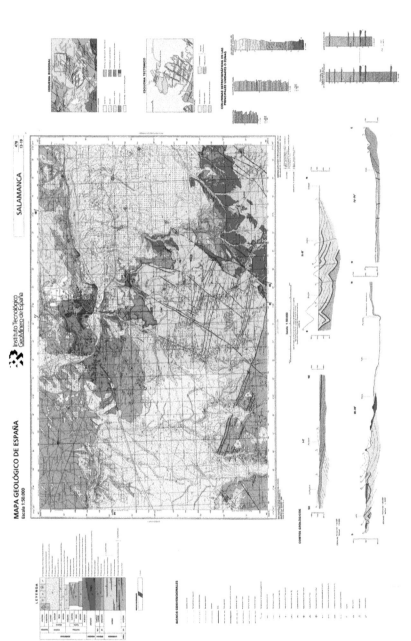

Figure 5.1a Geological map, at 1:50,000 scale, of the areas of Salamanca province, from where the most important stones for the construction of historical buildings were extracted. In this figure, most rocks are Tertiary sedimentary stones.

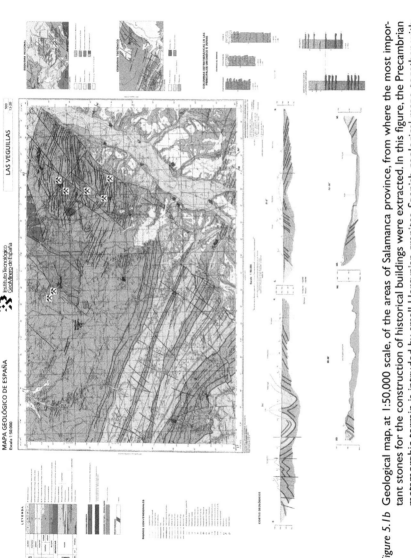

Figure 5.1b Geological map, at 1:50,000 scale, of the areas of Salamanca province, from where the most important stones for the construction of historical buildings were extracted. In this figure, the Precambrian metamorphic terrain is intruded by small Hercynian granites. Spanish geological maps, together with full geological descriptions, can be downloaded from the Spanish Geological Survey at http://info. igme.es/cartografiadigital/geologica/Magna50.aspx.

Stones for monuments and historic buildings were cut from large, quarried blocks that were sawn to the correct size and prepared for emplacement. The market for dimension stone has grown exponentially through the centuries, parallel to the improvement in construction techniques and stone-cutting technology. At present, buildings are constructed mainly of brick and concrete. Natural stones are used only as facings to cover façades, with slabs ranging from 2.5–5 cm thickness. The ability to economically cut large blocks into thin slabs and tiles to cover façades has made it feasible to use stone in commercial buildings and private homes. Even so, the stone should still meet specific requirements to be applied for specific uses.

Stones can be described in terms of technological characteristics. Some characteristics are more important than others, depending on whether they are intended for construction or ornamental use. These include mineralogical composition, chemical composition, density, water absorption, compressive strength, freeze-thaw, abrasion, flexural strength, etc. In the case of stones that are evaluated for restoration of historic buildings, the most important testing is listed in Table 5.1.

Some of the rocks that are found in historic buildings are not widely used commercially. Rather, they have been used only locally and sometimes only during a specific period. For some of these, there are no recorded technical descriptions, and, for some others, the characterisation is incomplete. Future work on them is needed.

To understand the quality of the stones regarding their technological characteristics, the latter are compared to the recommended values in published standards. The most useful are the European norms

Table 5.1 Recommended tests for dimension stones for the specific uses in exterior environments. All testing should follow a strict procedure in terms of international quality standards.

	Exterior paving	Exterior coating	Masonry use
Compressive strength	X		X
Flexural strength	X	X	
Water absorption	X	X	X
Capillary absorption	X		X
Freeze-thaw	X	X	X

Table 5.2 Requirements for the different tested parameters. European norm (EN) refers to natural stone in general, with a span of values. American norm (ASTM) refers to the different lithologies. In the table, only sandstone and granite are referred to, for simplicity.

Testing	EN requirements	ASTM requirements	
		Granite	Sandstone
Flexural strength (MPa)	9.80–19.61	> 8.27	> 2.41
Compressive strength (MPa)	39.22–147	> 131	> 28
Water absorption (atmospheric pressure) (%)	0.1–1	< 0.40	< 8
Density (kg/m³)	> 2600	> 2560	> 2000

set out by the European Committee for Standardization (EN) and the American Standards for Testing Materials (ASTM) for the conservation and restoration of architectural heritage, using the local, original stones (see Table 5.2). The British Standards Institute (BSI) also publishes codes of good practices for the installation of natural stone and specifications. In this book, the ASTM-recommended values have been used to describe the suitability of various stones, because the ASTM values distinguish between the different type of lithology (e.g. granite, marble, limestone, sandstone, serpentinites), while the European norms refer only to natural stones in general (EN, 2008).

Chapter 6

Stone testing

Each stone is suitable for a specific purpose. Suitability is determined by the careful study of the physical and mechanical characteristics, all of which are directly related to the mineralogy and texture of the stone. Mineral composition consists of essential (main constituents) and accessory (minor constituents) minerals but also secondary minerals, derived from the weathering of the first two. Analysis requires a detailed description (petrographic study) relating to both the macroscopic properties of the rock by visual evaluation of hand specimens, and the microscopic properties by a thin section evaluation using a petrographic microscope (ASTM, 2015a). The resulting petrographic description characterises the colour, fabric, mineralogy, grain size, cracks, cavities and evidence of weathering and alteration in both hand specimens and thin sections.

Porosity, either primary or secondary, generated from fractures, is also very important in the behaviour of the stone. It is an indication of the free intergranular or interstitial pore space in the material and can be estimated by carrying out a water absorption test (described later) or through specific testing by mercury (Hg) porosimetry (ASTM, 2004).

For our purpose, when determining the use of the stone in the restoration of historic buildings, there are a limited number of tests that need to be considered. The ASTM Standards used for the characterisation of the stones described in this book are as follows:

C 97 Test Methods for Absorption and Bulk Specific Gravity of Dimension Stone.
C 170 Test Method for Compressive Strength of Dimension Stone.
C 880 Test Method for Flexural Strength of Dimension Stone.

The following is a brief description of each tested property:

Water absorption. The water absorption coefficient is very important when the stone is going to be used in exterior environments, which are in contact with rain and humidity, and also with the upward percolation of underground water. The absorption coefficient measures the percentage in the weight of water absorbed by a dry stone sample. The tested sample has to have a very specific size according to the norm (ASTM in this book) to be representative. If the stone is to some extent anisotropic, the testing has to be done in two directions: parallel and perpendicular to the anisotropy.

The relationship between the absorption of water at atmospheric pressure, the density of the stone and the porosity is examined during the testing. There is an inverse relation among these, thus a smaller water absorption coefficient for a rock with high density and small porosity is good for the purpose of construction and restoration of heritage. A highly porous stone absorbs more water and is therefore more susceptible to alteration, weathering and deterioration.

Compressive strength. This is a very important property for the identification of suitable stone. A stone under water-saturated conditions will have a different value for this property than that of dry stone, which is important in relation to exterior uses, as we will see with some of the stones described in this book. In general terms, the water-saturated sample will have less strength than the dry sample. Again, the tested material has to have a very specific number of samples and size according to the norm in order to be representative (ASTM, 2017). If the stone has some anisotropy, the testing has to be done in two directions: parallel and perpendicular to the anisotropy. If the stone presents an anisotropy such as foliation, the compressive strength value will be higher when the principal stress is perpendicular to the foliation. This is very important because a foliated stone can be good for a specific use when placed in a particular orientation, but may fail if it is wrongly positioned.

Flexural strength. Flexural strength (ASTM, 2015b) is the ability of a stone (or any material) to withstand bending forces perpendicular to its longitudinal axis. The resulting stresses are a combination of compressive and tensile stresses. Sometimes, depending on the use of the stone, a flexural test is more appropriate. For example, for slabs and kerbs, this is a critical test. A higher flexural strength value indicates a higher bending strength. The required minimum values range from 2.8 MPa for low-density limestone to 10.3 MPa for granite.

Freeze-thaw. This parameter is very important when the stone is going to be used in a context where drastic temperature changes take place, either seasonally or between day and night, because of meteoric conditions such as rain or snow fall and freezing periods (ASTM, 2013). This is the case in Salamanca, where the continental climate changes temperature from very cold in the winter time (sometimes –10 °C) to very hot in the summer (sometimes above 40 °C).

Testing requires a number of samples to be dried in an oven, cooled then immersed in water. After this, the samples are placed in a tank inside a freezing chamber to reduce the temperature to –10 °C. After this, the sample is thawed. Several cycles are carried on, always following the specific norm. Finally, the previously explained strength testing is carried on the samples to see if there are any changes in variation. If the percentage difference of results is very high, the sample should not be used for specific uses in exteriors (like paving or cladding in humid and very cold geographic areas).

Capillary rise. Capillary rise is the main mechanism by which water penetrates into a building material from below. It is, by definition, the upward vertical movement of ground water through a permeable wall structure causing the appearance of rising damp in the structure. Building materials have pores of different shapes and different diameters, which means that they contain air voids that can become occupied by, and permit the flow of, water (Karagiannis *et al.*, 2016). This is one of the most important parameters to test when the stone is to be used for the foundations of a building, and it is essential if the porosity value obtained for the rock is higher than 1%. Water can reach a building material in several ways, e.g. driving rain, condensation of air humidity, runoff from roof and façade and/or capillary rise of ground water (Karoglou *et al.*, 2005, and references therein). Even so, building materials usually contain an amount of physically bound water without affecting their durability. But if the material's moisture content is above a certain percentage, the deterioration effect of moisture is activated, causing various physical, chemical and biological processes:

- Physical processes: water will dissolve and transport contaminants such as soluble salts. If wet, the material can become susceptible to damage because water expands when it freezes, thus a major decay phenomenon is the formation of ice in low temperatures, which may end up breaking the stone.

- Chemical processes: as water penetrates into a building material, salt crystallisation may occur at, or just under, the surface. Some salts are hygroscopic, facilitating water vapour absorption and, in many cases, causing further structural damages.
- Biological processes: moisture may stimulate the growth of bacteria, fungi, algae or even plants (Fig. 6.1) on the substrate with possible physical and chemical damage and also potential health risks.

Thus, the knowledge of the water movement within a building material is of great importance to determine the likelihood and mechanisms of degradation. The main decay mechanisms are these: hydrolysis, dissolution, hydration, oxidation, capillary rise, salt transfer and crystallisation, hygroscopicity and cycles of wetting/drying. The main types of consequent decay are these: spalling, peeling, delamination, blistering, shrinkage, cracking, crazing, irreversible expansion, embrittlement, strength loss, staining discolouration

Figure 6.1 Physical, chemical and biological processes affecting a historical building façade: Casa Lis in Salamanca.

Source: Picture by Vicente Sierra Puparelli.

Figure 6.2 Slabs of slate used to decrease the water absorption by capillarity in the lower part of a building in Salamanca constructed using Salamanca sandstone (opal-cemented conglomerate).

and bio-decay of building materials (ICOMOS, 2008). Potential sources of water are these: the ground, the environment (rain, sea, water vapour, etc.), possible water sewage leakages, use of water for the production of building materials and interventions with the use of extensive quantities of water and salts. The best way to "fight" moisture-related problems in buildings is the prevention of the entrance of moisture at the design and construction stages (Fig. 6.2). However, the elimination of the problem in existing structures, especially historical ones, is more problematic.

SEDIMENTARY STONES

Sedimentary stones are formed through the compacting of accumulations of grains or fragments of existing rock material which have been formed through weathering and transport by water, wind or ice. Following burial, they are liable to compact and cement through a physical or chemical process to form cohesive rocks.

These are the most common rocks in the buildings of Salamanca, due to the surrounding geology and the difficulties of transportation at the time that the first historic buildings were constructed. In the nearest surrounding area, Tertiary sandstones are widespread in the landscape. These stones have been used as construction materials through the centuries. Rocks from the Palaeocene and Eocene time periods were the preferred rocks for that purpose, but they are the most challenging stones in terms of conservation. Loss of material is the most common damage (ICOMOS, 2008)

METAMORPHIC STONES

Metamorphic stones result from the alteration of pre-existing rocks due to high temperatures and/or pressures. The pre-existing rocks may be igneous, sedimentary or previously metamorphosed rocks. The newly generated metamorphic rock has substantially the same composition, in terms of chemical elements, as the parent or original un-metamorphosed rock, but it is re-arranged into new minerals and textures.

These materials are not very common in the historic buildings of Salamanca, as they are not widespread in the surrounding geology, but they have otherwise played a very important part in the construction of buildings and structures. Although there are marble and recrystallised limestone quarries in nearby areas (for example, Casafranca, around 60 km south of the city), only the slates of Mozárbez and Pizarrales (a neighbourhood at the outskirts of the city of Salamanca) were used in the construction of the buildings of what is recognised as a World Heritage Site today. Recrystallised limestone and marble were probably not used because it was difficult to access their outcrops at that time. However, some structures and ornamental works in highly valued buildings in Salamanca were built out of very well-known imported metamorphic stones. For example, La Purísima, a church built during the 17th century, is a work of art, mixing sculpture, architecture and painting. In this building there is a mixture of stones from Salamanca and other parts of Europe, mainly Italy. Columns of Carrara marble with incrustations of Red Verona marble (in fact a limestone), Yellow Siena, Belgium Black and Green marble from Calabria (in fact a serpentinite) are found decorating this church and other historic buildings. Although these imported stones were used in the construction of very important buildings, these imported materials are not considered in this book as part of Salamanca's heritage built of stone.

IGNEOUS STONES

Around Salamanca we only find granitic igneous rocks. They are formed through the cooling and solidification of magma with high silica content deep within the Earth under high-pressure and slow-cooling conditions. Granitic rocks outcrop in several places in the province of Salamanca and began to be quarried for construction purposes when it became clear that their behaviour and resistance was much better than sedimentary rocks. Granite was used as the main construction material during Roman times, and in Salamanca it is possible to see some constructions from that period, of which the Roman bridge is the most striking example (Menéndez Bueyes, 2000–2001), but also the long Roman road connecting the north with the south of Iberia, passing by Salamanca (Vía de la Plata; see Chapter 3).

The quarries

Landscapes change because of dynamic interactions between natural and cultural forces in the environment. They are partly the consequence of continual reorganisations of land in order to better adapt uses and spatial layouts to changes in social demands. There is also a link between landscapes and the cultural events and traditions that make a mark on peoples' perceptions. This is the case with the quarry landscapes, which are not only culturally rich, but are also related to geological evidence and heritage (Douet, 2015). While the quarrying of stone is an old industry, it is still economically important. The natural stone is a high-value natural resource both for new construction and for the restoration of historic buildings.

Quarrying has always been connected to the development and evolution of the means of transport; thus, the use of local quarries to provide stone for the culturally important architectural heritage led to the city being designated a World Heritage Site by UNESCO. Even though some of these quarries have been closed for many years, due to their cultural value, they should be preserved as a natural resource for possible future use. Also, existing quarries should be kept open to provide stone for any future repairs to historic buildings. This good practice of safeguarding important sites has been implemented in some European countries where the value of the heritage built on stone is greatly appreciated. Such is the case of the Podpeč limestone in Slovenia; it is a dark grey limestone with white shells of fossil bivalves that Romans used to embellish the most important towns of the country. Today the Podpeč limestone quarry is protected as a natural treasure and only extraction of the small quantities required for the restoration of buildings is permitted (Kramar *et al.*, 2014). Podpeč limestone has also been recognised as GHSR (www.globalheritagestone.com) (see later sections).

HERITAGE STONES

The concepts of Global Heritage Stone Resource (GHSR) and Global Heritage Stone Province were developed by the International Union of Geological Sciences (IUGS) Heritage Stones Subcommission (HSS, previously the Heritage Stone Task Group, HSTG) in 2008. This proposed that all natural stones of international or national significance used in the construction of historic buildings and monuments over an extended period, preferably centuries, should be recognised through a designation because of the need for appropriate repair, maintenance and restoration of the cultural heritage, even if the quarries are not active anymore. The designation as GHSR is achieved only after a strict procedure defined by the HSS is followed (Marker, 2015) and if it is ratified by the IUGS Executive Committee.

Candidate stones have to meet several requirements. At least, the stones have to have a minimum historic use of at least 50 years, although centuries are preferred; they preferably have to have a wide-ranging geographic application, but can have a very local use if they are recognised as a cultural icon, including an association with the local, regional or national identity or a significant individual contribution to architecture; they have been used in the construction of significant public or industrial projects; there is ongoing availability of material for quarrying, even if the quarry is not still active; and their recognition as GHSR has potential benefits in the cultural, scientific, architectural, environmental and/or commercial arenas. This designation as GHSR has scientific value through the formalisation of the description of technical properties of stones as well as cultural value.

In 2017, the first stones received such a designation: Portland stone from the UK, Carrara marble from Italy, Podpeč limestone from Slovenia, Estremoz marble from Portugal, Petit granit from Belgium, Larvikite from Norway, Hallandia gneiss from Sweden and Villamayor sandstone from Spain. Other stones, such as Martinamor granite and Salamanca sandstone, are already on the list of candidates for consideration under the protocol and, if meeting all requirements, will achieve GHSR designation. Stones from the Salamanca area can also make a case for achieving the designation as Global Heritage Stone Province (Pereira and Cooper, 2015).

Heritage stones in Salamanca

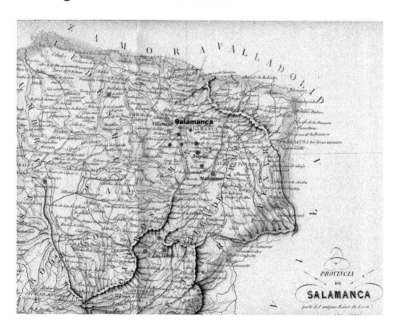

Figure 6.3 Distribution of historic quarries in Salamanca and the surrounding area. Brown dots: opal-cemented conglomerate; orange dots: Villamayor sandstone; red dots: granitic rocks (including vaugnerites); blue dots: slates.

Source: The map is a facsimile from Hevia (1860), reproduced courtesy of the county of Salamanca.

THE SANDSTONES

In this book we will learn about two types of "sandstone" that we can admire in Salamanca: Salamanca sandstone and Villamayor sandstone (Fig. 6.3). The first is, in fact, an opal-cemented conglomerate known locally as *Piedra Tosca*. It catches people's attention in historic buildings due to its reddish and whitish tones, which contrast sharply with the golden Villamayor sandstone.

Salamanca sandstone lies beneath the city and was discovered in the excavations carried out when the foundations were being laid. Subsequently recovered blocks were reused for backfilling trenches and the remaining sandstone was placed in the base levels of buildings. This stone is less porous than Villamayor sandstone, thus it better supports foundations than the golden stone.

Villamayor sandstone comes from the village of the same name. It is the most important stone used in the monuments of Salamanca, both for support and for aesthetic purposes. When it is quarried, it is very wet, which makes the task easier. Taking advantage of this feature, stonecutters hewed the stone in the quarry and then brought it to the city. Because it weighed less, the cost of transporting the stone was lower. Also, because of the ease of carving when wet, the golden stone acquired the beautiful shapes we can now admire in façades. As time passes, Villamayor sandstone loses moisture and becomes hard and robust.

This sandstone has high amounts of iron oxides in some parts, which produces a characteristic colour on the façades as evening falls and the light shines on the stone.

Villamayor sandstone

Villamayor sandstone is the most widely used stone in buildings of cultural significance in Salamanca (Fig. 6.4). It is also known as *Piedra Franca* (which can be translated as "Honest Stone" in English) or golden stone. At dawn these façades made of Villamayor sandstone seem to glow with a golden hue in the sun, a feature which appears in many literary works and from which the city takes its nickname: the Golden City (Stock, 2012). Villamayor sandstone is an arkosic sandstone of the Middle Eocene age. Several scientific papers have dealt with this sandstone due to the importance of the material both for construction and restoration of Salamanca's cultural heritage.

Figure 6.4a The historic building of the university. The University of Salamanca is
the oldest university in Spain and one of the oldest in Europe. The existing buildings date from between 1429 and 1520 and were restored
between 1767 and 1869 (Azofra and Pérez-Hernández, 2013). The
architects in charge of the construction and restoration were Alonso
Rodríguez, García de Quiñones and Secall Assión. The university building owes its most essential features to its intricate Plateresque façade
(also known as the Rich Façade), which is entirely constructed from
Villamayor sandstone, and the ornamental carvings of which depict
numerous scenes of a religious, mythological or symbolic nature. The
high capacity for water absorption, together with its porous network
and mineralogical composition, makes Villamayor sandstone a material
that can be easily worked and elaborately sculpted when wet. The sandstone rests on granite base stones from Martinamor. The door-frame of
the main entrance and the base, which are carved, are made out of granite from Los Santos. But those granites are in general a subsequent addition as the result of replacements during restorations. The university
was initially built completely of Villamayor sandstone and it was in 1932
that the replacements took place, as recorded in the photo archive of
Marburg (www.fotomarburg.de/). Not all the replacements were done
using Martinamor granite. Slabs of Los Santos granite are found replacing blocks of sandstone in the lower parts of adjacent buildings.

Source: Picture by Vicente Sierra Puparelli.

It is part of the Cabrerizos Sandstone Formation, constituting
one of four groups of sedimentary rocks (lithofacies) that vary in
character laterally. The Villamayor sandstone rests discordantly
on the Upper Cretaceous Salamanca Sandstone Formation (or

Figure 6.4b (Continued)

opal-cemented conglomerate). The basal deposits of the arkosic Villamayor sandstone are very porous, while the upper deposits are more silicate rich subarkoses, and metamorphic mineral grains are more abundant. Above, and discordant with them, are red conglomerates of Miocene age, which give a reddish tinge to the deposits of Villamayor sandstone beneath the disconformity (García-Talegón *et al.*, 2015, and references therein).

Villamayor sandstone is a feldspar-rich sandstone composed of 40–70% quartz, 10–30% feldspar, micas and an abundant clayey matrix. It can contain other minerals in small proportions, such as tourmaline. Feldspars are usually very altered, and the rock has high porosity (up to 30%) and high water absorption (up to 14%). The stone has variable resistance values, which considerably decrease when in a state of water saturation. And this is why Villamayor sandstone, once totally dried, increases in durability. Vielva (2001) found that compressive strength varied from 2.6 to 28.5MPa, depending on whether the sandstone was saturated or dry. Flexural strength also varied from 0.6 MPa when saturated to 2.6 MPa when dried. A complete characterisation of this stone can be found at the referred doctoral thesis (Vielva, 2001).

Table 6.1 Comparison of durability values of Salamanca sandstone with the ASTM recommendations for sandstones ASTM (2015c).

	Compressive strength (MPa)		Flexural strength (MPa)		Water absorption (%)	Capillary absorption coefficient (g/cm²s¹/² 10⁻³)		Freeze-thaw (%)
	Wet	Dry	Wet	Dry		Wet	Dry	
Villamayor sandstone*	2.6–28.5		0.6–2.6		< 14	70–85		5.7 (average)
ASTM specifications	> 28		> 2.4		8	n.a.		n.a.

*Ordaz (1983) and Vielva (2001). n.a. = not available.

From Table 6.1, it can be seen that Villamayor sandstone, when dry, has competitive values of compressive strength and flexural strength, compared to the ASTM recommended values. The maximum value of water absorption is far too high, compared to the standard. This issue has proved to be a problem. Also, the capillary absorption of this rock is very high, depending on the microporosity (Ordaz, 1983). The results clearly indicate that the rock is suitable for exterior use, but should be avoided for use in the base courses of buildings. Several examples can be found in Salamanca in historic buildings from the 12th century where the whole building, including base courses, was made out of Villamayor sandstone with consequent deterioration in the lower parts (Fig. 6.5). It has been recommended that, around those buildings, it is wise to construct a drainage area that will allow the stone to "breathe" and the underground water to evaporate instead of rising up through the sandstone lower ashlars, as it has been done in other places with buildings using stones of similar absorption characteristics (Fig. 6.6).

The quarries

The most important Villamayor sandstone quarries are located in the village of the same name, Villamayor de la Armuña, which is 5 km to the east of Salamanca (Figs. 6.3 and 6.7). Some quarries of this natural stone remain active because a local ordinance stipulates that it should be used for the construction and restoration-conservation work in Salamanca's historic city centre (Fig. 6.8).

Figure 6.5 San Marcos church, a Romanesque building from the 11th–12th centuries. This church was initially on the outskirts of the city. Completely built on Villamayor sandstone, the deterioration in the lower part of the building is clearly seen, with loss of material and colonisation by lichens due to humidity.

Figure 6.6 To illustrate an example of good practice: building made of limestone in a central street of Oxford. Limestone has the same water absorption problem as sandstone. However, some of the British buildings have partially neutralised the humidity effect by creating a drainage area around the building.

Figure 6.7 Sandstone quarry in Villamayor. Quarries of this stone are small and medium in size, depending on the extent of usable stone strata that appear during investigation. The village of Villamayor has increased in population in recent years, so the centre of the town cannot now be used for the extraction of stone. The largest volume of un-extracted stone is now towards the river outside of the village.

Figure 6.8 Construction of historic buildings used full blocks of sandstone. However, that approach is no longer possible for economic reasons. Modern buildings are constructed with bricks and concrete, and natural stone is used mostly for aesthetic reasons such as facings. To reduce the amount of stone used, a new methodology was introduced through the use of nails in the slabs of sandstone. These nails produce a thin oxide patina that increases their volume and, therefore, produces a better tightening tension. Nails are tied with wire to increase the strength of adherence to the concrete and plaster to cover the building façades.

Natural stone quarrying was one of the most important economic activities in this part of Spain. However, the recent economic crisis (mainly hitting the construction industry) caused the closure of many quarries and continues to put sandstone extraction at risk. There are also historic quarries which are very well preserved and from which the ashlars for the construction of significant buildings, such as the cathedrals of Salamanca, were quarried. Nowadays these historic quarries are a tourist and educational attraction, where traditional stone quarrying processes are demonstrated. Such activities also help preserve these old quarries.

As there are now other types of sandstone, with similar features but lower prices (e.g. Dorada Urbion and Beige Pinar from Spain; Bombay sandstone from India), Villamayor sandstone is no longer the source of wealth it once was, but it is still quarried. However, it is important to realise that although these other sandstones are

Figure 6.9 Villamayor's historic quarries are of great importance for teachers and researchers, as they contribute to the understanding of the historical importance of extraction activity, which was carried out when Salamanca's main historic architecture was built. The Old and the New Cathedrals of Salamanca are made completely out of stone blocks from these quarries.

somewhat similar in visual appearance, their physical and mechanical properties can greatly differ, which should be taken into account when identifying suitable materials for replacement of Villamayor sandstone (see www.tectonica-online.com/products/1826/).

There are also historic quarries (Fig. 6.9) which are very well preserved and from which the ashlars for the construction of significant buildings such as the cathedrals were quarried. Nowadays these historic quarries are a tourist and educational attraction, where traditional stone quarrying processes are demonstrated. Such activities also help preserve these old quarries.

Salamanca sandstone: opal-cemented conglomerate

The rock known as Salamanca sandstone is geologically, in fact, a conglomerate (Fig. 6.3). It is known locally as *Piedra Tosca* ("Rude Stone", to differentiate it from *Piedra Franca* ["Honest

Stone"] Villamayor sandstone). The name "Salamanca sandstone" derives from the geological Salamanca Sandstone Formation (Nespereira *et al.*, 2010, and references therein). This formation is a late Cretaceous–early Palaeocene–deposit consisting of siliciclastic sediments that accumulated in braided river systems. The formation presents two facies, one of which was intensively cemented, derived from silica exchange with intrasedimentary palaeosols (Blanco, 1991). The opal-cemented conglomerate has increased durability. Molina *et al.* (2009) concluded that the cementation process took place as a function of the porosity of the sediment as the previous palaeo-drainage evolved (Fig. 6.10).

Some physical and mechanical features of these stones have been published in several research papers that are listed in the bibliographic references (Nespereira *et al.*, 2010). Table 6.2 is a summary of the characterisation.

This stone has a high iron mineral content, which is responsible for the reddish tone which covers patches of the buildings in a heterogeneous way, due to colour variations in the wide spread of quarries in and around Salamanca. Unlike the Villamayor sandstone, the conglomerate has not been damaged as much by the passage of

Figure 6.10 An opal-cemented conglomerate appears with mixed red and white colours, depending on the outcrop. Opal cements the grains but also appears as silicified levels.

time and exposure to wet conditions when it was used on the lower parts of building façades. Nonetheless, many historical buildings have frequently required restoration-conservation work. The lack of professionalism in restoration, with undesirable practices such as the use of mortar and cement to cover the blocks, has accelerated stone deterioration (Figs. 6.11 and 6.12).

Table 6.2 Comparison of durability values of Salamanca sandstone with the ASTM recommendations for sandstones ASTM (2015c).

	Compressive strength (MPa)	Flexural strength (MPa)	Water absorption (%)	Capillary absorption coefficient ($g/cm^2 s^{1/2} 10^{-3}$)	Freeze-thaw (%)
Salamanca sandstone	15.8–177.6	n.a.	<14	n.a	n.a
ASTM specifications	>28	>2.4	8	n.a	n.a

n.a. = not available.

Figure 6.11 Deterioration in opal-cemented conglomerate accelerated by the mortar coverage used in a failed restoration attempt. The picture shows part of the façade of the Clerecía church, one of the most important monuments in Heritage Salamanca.

Figure 6.12 Montellano palace. Built during the 15th, 16th and 17th centuries, it was originally a palace before housing the religious order of the Barefoot Trinitarians. Today the building is a student hall of residence. A mixture of architectural styles, Gothic, Renaissance and Baroque styles can be identified combined in structural elements that were later altered. Inside the building there is a proto-Baroque two-storey cloister. The basement of the façade was made of the silica conglomerate (Salamanca sandstone), which, due to its deterioration over time, was replaced with granite with little regard for aesthetic and architectural styles. It is an open-and-shut case of a restoration failure, probably because of the lack of knowledge of the lithologies by the architects in charge of the restoration.

There is a wide range of values for the physical and chemical parameters due to variations of the stone in the different outcrops, mainly related to the degree of cementation. It also depends on the composition of the cement, as Nespereira *et al.* (2010) pointed out. Unfortunately, references lack information on flexural strength values. This parameter is very important when using the stone for paving. The maximum values of compressive strength can be positively compared to the standard specifications for sandstones. This was probably noted by past architects when choosing the material with which to build the foundations of the buildings of Salamanca.

Figure 6.13 One of the abandoned millstones, in the countryside 20 km to the south of Salamanca, made of opal-cemented conglomerate, extracted from a quarry nearby.

Opal-cemented conglomerate was a versatile rock that was used not only for construction. It is frequent to find in the field millstones that were used to grind cereal into flour in the 19th century (Fig. 6.13). In fact, the inhabitants of the area in the south of Salamanca knew the stone as a "millstone" (*piedra de moler*) (Madoz and Sagasti, 1845).

The quarries

Salamanca sandstone or opal conglomerate was extracted from quarries around and in the vicinity of Salamanca (Fig. 6.3). In fact, the city of Salamanca is built on top of this formation. None of these quarries are active except for occasional restoration activities on some historic buildings in the city centre. Yet there are many other historic quarries near the city where the extraction process can be observed. One of these historic quarries is also connected to an historic site in the province: Arapiles. In this village, which is 5 km to the south of Salamanca, there are two inselbergs consisting of this material (Fig. 6.14).

Figure 6.14 View of Arapiles from the road from Salamanca towards the south of Spain.

This was a strategic point where the Battle of Arapiles (Figs. 6.15a and 6.15b) took place in 1812 during the War of Independence against the French. It was here that the Duke of Wellington, commanding the Anglo-Portuguese army and helped by some Spaniards, defeated Napoleon's troops. As a result, these quarries are perfectly preserved, not because of the importance of the stone, but due to their historic relevance. It is an attraction, mainly for British tourists, as the terrain is perfectly preserved and details of the battle are very well explained (Pérez Delgado, 2002).

Today it is still possible to see some restoration of monuments and buildings using this conglomerate because, when a new building is constructed in Salamanca, the underlying conglomerate is dug during the building of foundations and is taken to a stone company outside of the city for preparation so that architects can use this stone when required. This action has to be applauded in terms of preserving architectural heritage (Fig. 6.16).

Another interesting restoration action is the use of aesthetically similar stones for replacement (Fig. 6.17), which could be a strategy to follow when the original stone is not available.

Figure 6.15a Memorial for the British soldiers who fell at the Battle of Arapiles. Every year, around 22 July, there is a celebration to commemorate the event, when many British people come to Arapiles to learn about the history.

Source: Picture by Vicente Sierra Puparelli.

Figure 6.15b One of the two Arapiles hills, with a monument dedicated to the battle. Indicators of the battle site are found in the area, and a small interpretation centre dedicated to the event can be visited in the village of Arapiles.

Source: Picture by Vicente Sierra Puparelli.

Figure 6.16 Lower part of San Martin church. Granite and opal-cemented conglomerate have been used to avoid further deterioration of this Romanesque religious building from the 12th century.

Figure 6.17 The building at the end of the street is the Faculty of Geography and History. It was built on the rest of the 16th-century Saint Pelayo college, totally destroyed during the French War of Independence. Inside the building it is possible to visit both the cloister of the previous college and also some pre-Roman remnants on top of which the college was built. The new building was constructed during the 1980s and won an architectural prize. The base course uses a red granite (Rojo Sayago, an episienite from Zamora province). The granite emulates the red opal-cemented conglomerate that was used on the complete façade of the neighbouring building.

MOZÁRBEZ

The greenish slate of Mozárbez was quarried in the village of the same name, which is 15 km to the south of the city (Fig. 6.3). Quarrying was a very important activity there, but nowadays the slate is used only for local construction in Mozárbez village. It was initially used together with other types of slate from the city (specifically from the *Barrio de Pizarrales*, "Slates neighbourhood", as it is called nowadays). This slate provided stability for the other stone blocks, mainly the sedimentary ones, as *calzo*, and stopped water absorption through the walls (see earlier).

Mozárbez slate, which was of higher quality than its equivalent from Pizarrales, was used in the construction of remarkable buildings such as the vault of Pozo de Nieve ("Well of Snow"; Fig. 6.18),

Figure 6.18 Bottom of Pozo de Nieve ("Well of Snow"), made out of Mozárbez slate during the 18th century. It is more than 7 m deep. Nowadays the vault is whitewashed to add a touch of light to the most interesting areas of the construction.

Source: Picture by Vicente Sierra Puparelli.

which originally belonged to a convent of the Barefoot Carmelites of Saint Andrew in the 18th century. It was used to store the snow (converted into ice) from the mountains around Salamanca and to provide the city with ice blocks. Mozarbez slate is also found in the flooring, roofs and balconies of historic buildings such as those in the Plaza Mayor (the main square of the city) and in the carving of identifying plaques, for example, that of the Royal Pavilion Plaza Mayor. But slates were used also in combination with other stones, mainly granites, to make a checkerboard-like pattern in the floors of some major buildings (e.g. the chapel at the Colegio Mayor Fonseca, also called the Irish College, founded in 1592 to house students from Ireland persecuted for being Catholics).

Because the slates of Salamanca are not used in the contemporaneous market for construction materials, there are no available data on their physical and mechanical characteristics. It has to be taken into account that the most appreciated slates in the market are those suitable for roofing, which surely have physical properties that differ greatly from the slates from Salamanca.

THE GRANITES

In the historic buildings of Salamanca, granite is the most widely used stone in the lower parts of the construction. Transporting granite involved a great deal of effort, using animal-drawn carts to bring the stone from the quarry to the construction site, so architects must have been well aware of the advantages of this material when they decided to start using it. The stone was carried mainly from Martinamor, Calzadilla del Campo (Ledesma) and Los Santos, although there is also evidence of other important quarries: Villavieja, Sorihuela, Vitigudino, La Magdalena and Linares, all of which are in the province of Salamanca (Fig. 6.3). Los Santos and Martinamor granites have been the most frequently used, first in construction and later in restoration-conservation activities, mainly in the historic buildings of the city centre. These two stones have been used in the architectural history of Salamanca from the 16th to the 20th century, as they were the most easily accessible via the means of transport at that time.

Since different materials appear together in one single building, the historic city centre of Salamanca reflects the use of the province's different materials through history and with the development of roads and transport. Some buildings of historic architectural interest are true patchworks of different materials that not only break up the aesthetic appearance of that heritage, but sometimes also endanger the structural integrity of the building due to the likelihood that the various stones deteriorate at different rates.

All granites described in this book have similar durability properties, but their large-scale appearance can be very different, so it is not recommended to use them indiscriminately for the restoration of historic buildings.

The fact that all granites used in façades many centuries ago have only sustained minor damage today confirm that it was an excellent idea to use these materials on the lower parts of buildings in place of sandstone when the quarries near the city ran out of this kind of stone. Where aesthetically acceptable, the use of these materials in restoration could boost sustainability, not only of the conservation of the architectural heritage of the province of Salamanca, but also of the area's ailing ornamental stone industry. However, more detailed and specific research should be done on the physical and mechanical features of all the stones, as we lack important data about some of them and others have given controversial results with regard to resistance and durability parameters. When stones have given good market results, complete characterisation of properties can be found even on the quarrying companies' websites, although careful scrutiny has to be given on the characterisation that is offered by private stone companies. But because some of the stones have not been used for recent construction, like Martinamor granite (which ceased to be used as a construction material in the middle of the 20th century), there are no data on their physical and mechanical behaviour, except in some specific research publications.

Usually only one type of granite was used for the base of buildings, but sometimes there appears to be a wider range of materials in the lower parts of the façades. This may be due to the inadequacies of some descriptions. For instance, we can often find historical literature referring to *Piedra Pajarilla* (Martinamor granite) as a construction material in a specific building while, in fact, it can be observed that the actual natural stone used matches other types of rock, frequently vaugnerites or granite from Los Santos. This sort of misinterpretation is very common in other parts of the area as architects and quarrymen used to give local names to stones. Just as in the province of Salamanca, all granites and granitoids were called *Piedra Pajarilla*; in the province of Ávila and in some parts of Extremadura, people use the term *Piedra Berroqueña* (Berroqueña stone) to refer to any type of granitic rock. Similar use of local naming is known also in other parts of the world. Architects selected the best material to build magnificent constructions like palaces, castles and cathedrals. A large volume of stone was needed for the purpose, but also a durable and easy-to-work material was preferred. A well-known case is found in France, where huge amounts of the pale-coloured Caen limestone was extracted for important buildings in France and England. This stone was known as "freestone", in contrast to "roughstone" (Pereira and Marker, 2016a). The original French *Pierre de Taille* term referred originally just to the limestones extracted at Caen, but, over time, the original meaning was expanded to include other natural stones with similar colouration and ease of carving. Notably this included many sandstones that were used in adjacent countries such as Spain, including Villamayor sandstone from Salamanca, as mentioned earlier. Another case that can lead to potential confusion related to the use of local names comes from the use of *Piedra Dorada* (golden stone) for limestones from Andalucía (Spain). This stone was used to build the architectonic heritage of places such as Ubeda and Baeza, which have been World Heritage Sites since 2003. In fact, Baeza is called the "Andalucian Salamanca" for the similarities in several features, including the colour of the buildings. But the big difference is that the stone in Salamanca is a detrital sedimentary stone – a sandstone – while the stone in Baeza is a chemical sedimentary stone – a limestone. Some mistakes were found in the promotion of the latter (e.g. www.rutasyleyendas.com/ruta23_ubeda/ ruta23.htm), where the author of the page claimed to have seen the golden limestone in "his dear Salamanca". This mistake has already been reported for correction (and has been corrected by the author of the web page, with acknowledgements). Another common link between the two places were the consequences of the terrible Lisbon earthquake in 1755 that severely affected the buildings. The effects can be traced in Salamanca's historical buildings, like the still tilted tower of the New Cathedral, and the old university building of Baeza, where it is possible to see the different cracks affecting the façade (Fig. 6.19). If a restoration has to take place, promoters should make sure to use the original limestone, not the similarly named "golden stone" from Villamayor.

In the following pages there is a description of granites that have been identified in several buildings: Martinamor granite or *Piedra*

Figure 6.19 Façade of the "Old University" of Baeza, dating from the 16th century and built of limestone. Open cracks in the middle and left superior parts of the façade are due to the 1755 Lisbon earthquake.

Pajarilla, Los Santos granite and Sorihuela granite (a detailed description of each of them is given in later subsections). Some research also describes base courses in buildings made out of vaugnerites that are called "granitoids" in some scientific publications. Other studies erroneously refer to that latter stone as *Piedra Pajarilla*. An explanation for this is given in the section referring to vaugnerites.

It has not been possible to make a definitive classification of all the granites and granitoids observed in the buildings of Salamanca, although granites from Vitigudino and from other parts of the province are mentioned in some technical reports and historic documentation as being used for the construction of paving and for different structural parts of historic buildings.

Sometimes it is difficult in Salamanca to find buildings that are homogeneous in terms of construction material. According to some research, more than the 25% of the buildings studied have a mixture of stones in their lower part, in comparison to the percentage of buildings in which the construction or restoration only used two or three different granites (González Neila *et al.*, 2017).

Martinamor granite

Juan de Álava was the driving force behind the use of this granite, which started to be quarried at the beginning of the 16th century. It is found in the ashlars and structures of the most important buildings of that time. In the 18th century it was also used to pave squares, streets and cloisters. During the 19th and the 20th centuries, Martinamor granite began to be used in the construction of more functional buildings.

Martinamor granite is light in colour and can be coarse or fine-grained, depending on the variety chosen (Pereira and Cooper, 2013). Its main feature is the presence of elongated tourmaline aggregates that accumulated in different patches in the stone. These aggregates are what give this stone its local name *Piedra Pajarilla* ("Little Bird Stone") as the aggregate texture can sometimes be imagined as resembling the shapes of flying birds (Fig. 6.20).

The second phase of the Hercynian orogeny produced the prominent subhorizontal foliation of the granite, and this is why the physical and mechanical characteristics of the granite differ, being better when the rock is tested perpendicular to the foliation. According to the ICOMOS definition (2008), Martinamor granite can occasionally show some alterations such as disintegration into sand, surface shedding and stone breakages due to these effects (Fig. 6.21).

Figure 6.20 Outcrop of Martinamor granite, where the characteristic tourmaline aggregates populate the face of the stone, like "birds flying in the sky".

Figure 6.21 Descaling in a façade of a historical building in Salamanca. This damage is due to the frosting-defrosting periods in the city. It may affect to the aesthetics of the building, but because, at the construction time, huge blocks of the stone were used for the building, this damage will not affect the structure of the building, at least for some more centuries.

Some data about the resistance of granite can be found in the bibliographic references (Pereira *et al.*, 2015) but research is still being done by the author of this book in order to promote the cultural and architectural value of this stone and its appropriate use for the proper restoration of Salamanca's historic buildings (Fig. 6.22a). Recent studies have shown that, depending on the quarry, the granite shows different physical and mechanical characteristics, sometimes not reaching the minimum required values for the main features such as water absorption or capillarity absorption, but

Figure 6.22a San Esteban convent. This huge complex was built between 1524 and 1610. Combining Baroque and Gothic elements, its initial construction phase was in the Plateresque style. Its impressive sculpted main door is protected by a great arch. Inside there is a bright dome and three cloisters. This building is related to several historic events, as well as to the patronage of the family Álvarez de Toledo. The stone base of the main entrance is made out of Martinamor granite, which continues along the main façade. The flagstones of this façade are homogeneous in terms of size, colour and texture. On its north side, the building is completely made out of Villamayor sandstone, which has severely deteriorated at the base due to the stone's high absorption index and because the building is surrounded by grass, which is continuously watered (Fig. 6.22b). The ten columns at the entrance to the convent are completely made out of individual blocks of vaugnerites.

Source: Picture by Vicente Sierra Puparelli.

Figure 6.22b North façade of San Esteban church.

in other sites the values significantly exceed the required values. Regarding compressive and flexural strength, all of the samples can be positively compared to the ASTM standard recommendations (Table 6.3).

The quarries

Martinamor granite, known as *Piedra Pajarilla* (Madoz and Sagasti, 1845; Pereira *et al.*, 2015), was quarried for centuries in the village of the same name, which is 15 km south of Salamanca (Fig. 6.3). Extraction ceased in the mid-20th century, and nowadays the quarries are abandoned and are part of a livestock farm. Despite that, the quarries are perfectly preserved: there are even granite blocks remaining in the quarry that are ready to use. This highlights the legacy that can be seen in different quarries. Although only two quarries had previously been documented (Madoz and Sagasti, 1845; López Plaza *et al.*, 2009), the field work carried out by the author of this book has located many different places (Fig. 6.23)

Table 6.3 Comparison of durability values of Martinamor granite with the ASTM recommendations for granites (Pereira, in prep.) ASTM (2015d). Martinamor granite is highly anisotropic, so testing was done in both stress directions, parallel and perpendicular to the anisotropy. The best values for all parameters were always achieved when the testing was done perpendicular to the anisotropy. Therefore, this must be the right way of placing the blocks of this granite for construction and restoration.

	Compressive strength (MPa)	Flexural strength (MPa)	Water absorption (%)	Capillary absorption coefficient $(g/cm^2s^{1/2}10^{-3})$	Freeze-thaw (%)
Martinamor granite	161–252	12.7–25.3	0.9–0.24	0.88–3.39	0.011–0.04
ASTM specifications	> 131	> 8.27	0.40	n.a.	n.a

n.a = not available.

Figure 6.23 Martinamor was part of an area also very rich in tungsten and tin mineralisation. The relationship between tourmalines from Variscan granitic rocks and mineralisation of tungsten, tin and gold has been described in many publications (e.g. Neiva *et al.*, 2007). Many small abandoned mines in the shape of open pits are found in the field, which are also part of the mining heritage of the area. Tungsten mining was a very important activity in the mid-1900s.

Source: Picture from Google Earth, modified by author.

where one can observe the process whereby the blocks were separated through the joints, with the help of triangular wedges to horizontally and vertically separate beds of stone into blocks (Figs. 6.24 and 6.25). So far, these quarries are well preserved even though protection measures have not been made (Fig. 6.26).

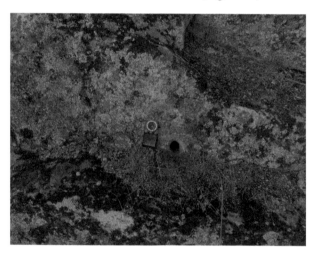

Figure 6.24 Horizontal wedge marks from the extraction of blocks of Martinamor granite.

Figure 6.25 Vertical, triangular wedge marks from the extraction of blocks of Martinamor granite.

Figure 6.26 Abandoned blocks of Martinamor granite in the quarry, ready to be used, if needed for replacement of damaged blocks in historic buildings. Blocks were manually extracted and shaped for use in construction.

There are many vestiges of earlier quarrying methods in many other outcrops around the area of Martinamor that are similar to those observed in ancient Egypt and prehistoric sites in Europe and South America (www.ancient-wisdom.com/quarrymarks.htm). These marks are due to the tools used to separate the blocks, both in the vertical and in the horizontal planes.

Los Santos granite

Los Santos granite is a typical representative of granitic rocks in the area of Salamanca, but also in the Spanish Central System. These granites are produced by partial melting of a metasedimentary protolith, the Schist-Graywacke complex (Pereira and Rodríguez-Alonso, 2000, and references therein).

With regard to its petrographic and geochemical features, Los Santos granite is actually a granodiorite. It is of light-grey colour and its quartz and feldspar crystals can be easily observed, giving

Table 6.4 Comparison of durability values of Los Santos granite with the ASTM recommendations for granites.

	Compressive strength (MPa)	Flexural strength (MPa)	Water absorption (%)	Capillary absorption coefficient (g/cm²s^{1/2}10^{-3})	Freeze-thaw (%)
Los Santos granite*	135	11.02	0.3	n.a.	n.a.
ASTM specifications	> 131	> 8.27	0.40	n.a	n.a.

*García de los Ríos Cobo and Báez Mezquita (2001). n.a. = not available.

the rock a phaneritic, porphyritic texture. These big feldspar crystals give this granite its character ("horse's teeth", in local terminology). The granite has imperfections in the form of microgranular enclaves. Apart from quartz and feldspar, Los Santos granite is also composed of plagioclase, muscovite and biotite (the last two make up 2.1% and 15%, respectively, which, when observed, appear as small crystals). It also contains accessory minerals such as cordierite, zircon and apatite. The physical and mechanical features of Los Santos granite are described in some scientific local publications (García de los Ríos Cobo and Báez Mezquita, 2001; López Plaza *et al.*, 2009). This granite, as with most granites from the region affected by the second phase of Hercynian age, is anisotropic, but published data does not distinguish between different stress directions for tests.

From the published values it is demonstrated that this granite, if properly selected, is a competitive construction stone which was known to the Romans who used it to build several constructions, including the Roman road that was also used to bring the blocks of granite to the emerging city.

This granite was next used at the beginning of the 16th century. Today it is the most widely used granitic material for new construction and restoration activities, although some poor choices have resulted in negative aesthetic effects (Figs. 6.27, 6.28 and 6.29).

The Casa Lis is the only building in Salamanca with an early 20th-century façade. The construction started during the last years of the 19th century and was finished in 1905, as stated in an inscription above the northern front door.

Figure 6.27 The Plaza Mayor is the most emblematic monument of the city. It was built between 1729 and 1755. Construction was started by Alberto de Churriguera and finished by Andrés García de Quiñones. It is considered a remarkable work of Baroque civil architecture in Spain. It has been renovated on several occasions using local stones, but also stones from other areas, such as Rojo Sayago granite, from the province of Zamora. The intention was to create a contrast with the grey granitic materials from Salamanca used in the paving of the Plaza Mayor. Villamayor sandstone ashlars on the façades rest on granite columns. Originally, the granitic material was the vaugnerite from Ledesma, but those responsible for restoration used granite from Los Santos which, being lighter in colour, disrupted the aesthetic appearance. A similar mistake has occurred when replacing damaged slabs of Martinamor granite by slabs of Los Santos granite, such as one major error in restoration that can be observed at one of the most beautiful buildings of Salamanca, the Casa Lis.

Figure 6.28 South façade of Casa Lis. The stairs go up to the central and main part of the building, with steps made originally with Martinamor granite. The balustrade was originally made of Villamayor sandstone, but extreme weathering caused the need to replace it. Unfortunately, the restoration was not done as carefully as it should have been.

Source: Picture by Vicente Sierra Puparelli.

Figure 6.29 The staircase, made of granite, climbs the south façade of the house, cutting across the great height from almost the river bank to the main living area of the house. This was evidence of the caprice of a rich man of the time. The staircase was originally built with Martinamor granite. Subsequent restorations have partially replaced them with granite steps from Los Santos. To the right of the staircase an efficient drainage system, made of small diameter holes with low slope, has been installed, preventing further deterioration of the façade. Due to the position of the house in a margin of the river, water absorption by capillary action is a danger both for the aesthetics of the façade and for the structural stability.

Source: Picture by Vicente Sierra Puparelli.

By then, the society in Salamanca was very provincial. Most people were farmers and only upper and lower classes could be distinguished. No middle class existed. This palace was the whimsical desire of a rich entrepreneur in the leather-tanning industry: Miguel de Lis. He was a modern man for those times, used to travelling to Paris, Berlin, Brussels and Vienna, and he wished to import the modern spirit into his parochial city. The architect Joaquín de Vargas was hired to create a modernist building using iron, ceramic, glass and brick but also natural stone (e.g. marble and granite). Joaquín de Vargas's Casa Lis in Salamanca could be compared in spirit to the well-known buildings of Gaudí in Barcelona.

To build the Casa Lis, part of the old wall surrounding Salamanca had to be pulled down, because Miguel de Lis wanted views of the river and the Roman bridge. This was only possible because of the status of the future owner of the house. This action could never have taken place today. But the destruction of part of one monument gave rise to the construction of another one. Miguel de Lis lived in the house with his family, but they had to sell the house in 1917. The new owner sold it again during the 1940s, and it was rented to a priestly congregation as a retirement house. In 1970 it was sold again, but nobody occupied it and the house started to deteriorate. At the end of the 1970s the house was a complete ruin. It had to be expropriated by the mayor of the city, Jesús Málaga. The restoration of the building was commenced by the architect Fernando Pulín. In 1995, a coloured, leaded glass representing the "Night" closed the central patio. Manuel Ramos Andrade, an antiquary living in Barcelona but born in Navasfrías (a small village southwest of Salamanca), donated his complete collection, including furniture, to Casa Lis under three conditions: the collection had to remain as a whole, it had to remain in Casa Lis and the benefits had to go to the elders and children of his birth village, Navasfrías. Thus, on 6th April 1995, one of the most important Art Nouveau and Art Déco museums in Europe opened to the public.

Three main features can be observed in the building: the north façade, the south façade and the central patio.

The north façade preserves the highest modernist style. It is made up by a façade, patio and gate, all of great simplicity. Many Art Nouveau details can be observed, like the floral and aquatic ornaments highlighted in relief on the door (Pereira and Pérez Castro, 2017).

The south façade concentrates the major artistic strength, mixing classicism and modernism. It was considered architecturally audacious for the time. It consists of a central structure, open at the lower level but closed at the upper level by an iron network supporting the stained glass. The latter was a new addition to add colour in 1992 during the restoration of the house. The original glass was not coloured. The iron introduced an innovative aspect in the building because it did not have any structural function, only ornamental.

The façade below and the balustrade of the stairs were made of Villamayor sandstone. Because of the heavy exposure to the extreme climate of Salamanca, being adjacent to the Tormes river, this stone became seriously deteriorated and a recent restoration took place, replacing most part of the balustrade with yellowish concrete. This has a very negative visual impact and is not acceptable in terms of art and beauty.

The stairs also suffered an unfortunate action when the Martinamor granite steps were replaced with Los Santos granite slabs, which resulted in another negative impact for the architectonic heritage of Salamanca (Figs. 6.29 and 6.30).

The central patio was originally open with different rooms opening from it. This structure is reminiscent of Andalusian patios and in fact it reflected the Andalusian origin of the first owner. The decoration in the corridors has been preserved. The floor is made of imported marble.

The museum Casa Lis holds several art collections, but the most spectacular is a collection of 2400 chryselephantines. These are fairly small sculptures (between 15 and 50 cm high, including the pedestal), common in the Art Nouveau period, representing human figures in ivory, gold, bronze, silver and natural stone and seeking to follow an Ancient Greek influence to represent their divinities. There is much published work on the figures but nothing on the pedestals, even though these are fundamental to the appearance of

Figure 6.30 Stairs of the south façade of Casa Lis. The original stone was Martinamor granite. The replacement of steps has been done with Los Santos granite, with a consequent negative aesthetic effect.

the works of art and are, sometimes, part of the figurines. That is unfortunate because the Art Déco and Art Nouveau artists were keen to find the perfect match between the stone and the sculpture. For instance, the colour of the chryselephantine's clothing sometimes resembles the colour of the stone of the pedestal. Almost invariably, pedestals are described as "marble" while, to an expert eye, it is clear that they are made of different kinds of minerals and rocks: onyx, serpentinite, limestone or even a schist with chiastolite crystals. Pereira and Pérez Castro (2017) described the pedestals of the most appealing pieces. Those authors consider this and other museums as good places for outreach activities to learn about the significance of minerals and stones in art.

The quarries

Los Santos granite is extracted from quarries that are located 50 km south of Salamanca (Fig. 6.3). There, granite is quarried, and then blocks and slabs are worked in adjacent factories (Figs. 6.31a and 6.31b). The stone leaves the factories ready to be used in buildings and structures. It is the most important quarrying activity in the province (Fig. 6.32), with several active quarries nearby the village

Figure 6.31a Commercial blocks in the open pit quarry of Los Santos. The blocks are sold as raw blocks or cut into slabs for emplacement.

Figure 6.31b Slabs of Los Santos granite, ready to be sold. The slabs and other granite products are manufactured adjacent to the quarry, where the company utilises the most up-to-date machinery.

Figure 6.32 The "Granite Thematic Park" in Los Santos. A few years ago, in an attempt to honour the stonemasons and the history of the stone as construction material, the city council of Los Santos created this original "park" with blocks of granite. Panels explain the quarrying history, the construction of Roman roads and the modern granite extraction industry and some traditional quarry tools accompany the display.

of Los Santos. This granite is now the most often-used stone used for the construction and restoration of buildings in the city centre of Salamanca. Consequently, several aesthetic mistakes have occurred in the restoration of buildings where Martinamor granite was originally used, due to the obvious differences between the granitic materials. This book shows some, but many can be observed during a tour around the city.

Sorihuela granite

Sorihuela granite has been used in modern conservation-restoration works on several buildings in the historic city centre of Salamanca, where the base course was not initially made of granite. There is evidence that this material had been used in the restoration of historic buildings in Salamanca from the end of the 19th century and the beginning of the 20th century (personal communication with the quarry workers) (Fig. 6.33).

This granite is extracted from family-owned quarries close to the village of Sorihuela, 60 km south of Salamanca. Its architectural purpose often matches that of Los Santos granite. Sorihuela granite differs slightly from Los Santos granite both in its colour, which is light grey and slightly blueish, and in its medium-grained crystal size, which is very homogeneous. It has the same imperfections as Los Santos stone, such as microgranular enclaves (*gabarros*, in the local language) and "horse teeth" feldspars, although not as many (Fig. 6.34) It is mainly composed of quartz, potassium feldspar, plagioclase and biotite. Its most common accessory minerals are apatite, zircon and opaque minerals. Its secondary minerals are sericite, muscovite and chlorite, derived from the alteration of feldspar and biotite.

Because of the small-scale extraction, far from the large commercial distribution centres that process the Los Santos granite, there are no current data available on the physical and mechanical characteristics. However, this granite is geologically very similar to Los Santos granite, affected also by the second phase of the Hercynian orogeny that produced subhorizontal structures and foliations, so it is reasonable to assume that the technical characteristics are likely to be similar.

The quarries

Granite quarrying is a long-standing tradition in Sorihuela (Fig. 6.34). Indeed, quarries were passed down through families for many centuries. The market for this stone mainly remains within

Figure 6.33 Salina's palace. The construction of this building started at the beginning of the 16th century but was not completed until 1881. At present it houses the county council offices of the Province of Salamanca. Notable architects such as Gil de Ontañón and Joaquín Casal worked on its construction. Some of its architectural features demonstrate the transition from the Gothic to the Plateresque style. Its basement is made of granite from Sorihuela, which was also used for the columns and some of the stairs, in combination with vaugnerite and granite from Martinamor. The steps that give access to the courtyard and the flooring were built with Sorihuela granite. The basement of the courtyard, made out of Martinamor granite, is very homogeneous. This same type of granite was used for the base of the columns and the stairs to the lower level.

Source: Picture by Vicente Sierra Puparelli.

Figure 6.34 Macroscopic view of Sorihuela granite. A few feldspar phenocrysts and the microgranular enclaves or *gabarros* are typical of this stone.

the limits of Salamanca province, but small quantities also go to the central and northern of Spain. Nowadays most of the product extracted from the quarry is used for cladding façades.

Villavieja granite

Villavieja granite outcrops in the Central-Iberian Zone (Díez Balda *et al.*, 1990). It is a pluton of Hercynian age, cutting the metasediments of Cambrian-Precambrian age (the Schist-Graywacke complex, Pereira and Rodríguez-Alonso, 2000). The pluton has a small subsurface dimension (around 58 km²). It is a yellowish, coarse-grained porphyritic rock, with feldspar megacrysts in a matrix of quartz, plagioclase, biotite and muscovite, with accessories such as cordierite, andalucite, sillimanite, apatite, zircon, tourmaline and some metallic minerals.

Villavieja granite has been used in various buildings and construction works in Salamanca. Its most prominent example is the paving of the Plaza Mayor (although some of the original granite has been

Figure 6.35 The Central Market in Salamanca. This modernist building was constructed between 1899 and 1907 by Joaquín de Vargas y Aguirre. The façade is a combination of Villavieja granite, iron columns, bricks and large glass windows, which are characteristic of this style.

Source: Picture by Vicente Sierra Puparelli.

Table 6.5 Comparison of durability values of Villavieja granite with the ASTM recommendations for granites.

	Compressive strength (MPa)	Flexural strength (MPa)	Water absorption (%)	Capillary absorption coefficient ($g/cm^2s^{1/2}10^{-3}$)	Freeze-thaw (%)
Villavieja granite*	135	14.3	-	n.a.	n.a.
ASTM specifications	> 131	> 8.27	0.40	n.a.	n.a.

*Data from PINACAL (www.pinacal.es/datos_tecnicos/variedad_info.asp?var=Granito%20VillaVieja&tipo=).

replaced over time) and in the Central Market's stone base: a modernist building (Fig. 6.35) that was designed by Joaquín de Vargas y Aguirre, an architect from Jerez, who also designed the Casa Lis.

At present, the author of this book is carrying out extensive research on this granite to fully characterise it (Table 6.5). Also, a radiological

Figure 6.36a View of one of the open pit granite quarries in Villavieja.

Figure 6.36b Pegmatitic variety of Villavieja granite. Both varieties are commercialised under the same name of Villavieja granite, which could also result in an unwanted aesthetic effect.

study is being carried out, as this granite was "demonised" for some time due to its proximity to uranium mines. However, preliminary results show that radon exhalation from this granite is below the recommended values for health safety (Pereira *et al.*, in prep.).

The quarries

Villavieja de Yeltes (Fig. 6.3) is one of the areas in Salamanca with the strongest tradition of quarrying. Well-known master stonemasons came from this village and took part in architectural works of great importance nationwide. Granite quarries in Villavieja have supplied both the city of Salamanca and the wider province with materials for the construction of dwellings, such as masonry blocks and slabs for paving. At present, there are two important, active, open pit quarries from which stones are quarried and then prepared in a factory in the same village (Figs. 6.36a and 6.36b).

LEDESMA VAUGNERITES

Vaugnerites are medium-coarse-grained rocks and consist of quartz, plagioclase and potassium feldspar as essential minerals, as well as biotite and amphibole as the main accessory minerals along with apatite. Their most relevant feature is the typical vaugneritic texture, which is manifested by large biotite crystals (Fig. 6.37). This is probably why it was mistakenly referred to as *Piedra Pajarilla* in some historical literature (Madoz and Sagasti, 1845), which has led to the misuse of this stone when replacements were needed in the restoration of some historic buildings.

Vaugnerites were quarried in Calzadilla de Campo, close to Ledesma (Fig. 6.3). Ledesma is a mediaeval village that has recently been declared as one of the "most beautiful villages in Spain" (Los Pueblos Más Bonitos de España, 2018). Declared a historic and artistic complex in 1975, it includes magnificent monuments and buildings made out of this stone, constructed between the 15th and 19th centuries.

Because of confusion with *Piedra Pajarilla*, this stone has been misunderstood in restorations since it is notably darker than other granites (see earlier for the description of the misleading restoration of columns in the Plaza Mayor of Salamanca, where the vaugnerite was replaced by the much lighter Los Santos granite).

While vaugnerites were formerly used in the construction of major buildings of Salamanca and on a large scale in Ledesma, nowadays they are only used for the construction and restoration in Ledesma's historical centre.

Figure 6.37 Mesoscopic aspect of the vaugnerite, in which big dark flakes of biotite represent the characteristic vaugneritic texture (Le Maitre, 2002).

López Plaza *et al.* (2007a) describe some of the physical and mechanical characteristics. They are dense rocks, with low porosity and low water absorption coefficient, which makes them ideal for the construction of the foundations of the buildings. The clever architects of the time chose them to build magnificent buildings of Salamanca such as San Esteban convent and Calatrava school (Fig. 6.38).

The base course of this building is made of Ledesma vaugnerite. It is a continuous frieze with similar-sized slabs but of different heights. This is the only building in Salamanca with these features, as the slabs are homogeneous in colour and texture. It is in a superb state of preservation. The upper part of the stone basement and the base of the frames and columns are carved. The outer stairs and slabs are made of Martinamor granite.

Figure 6.38 Calatrava school, also called Immaculate Conception school, was built in 1717 for the Order of Calatrava. This Baroque building is the greatest expression of this architectural style in the city of Salamanca. Its architects were Joaquín de Churriguera and Jerónimo García de Quiñones. It consists of a central courtyard with four towers and a fabulous façade emphasised by a staircase.

Figure 6.39 View of the area from which vaugnerite was historically extracted, on the outskirts of the village of Ledesma.

Source: Pereira *et al.* (2018).

The quarries

Vaugnerites were extracted from the quarries in Calzadilla del Campo, close to Ledesma, 34 km west of Salamanca (Fig. 6.3). These quarries are no longer active, except for occasional restoration activities, but evidence of old quarrying can be appreciated in the vicinity of the village (Fig. 6.39).

Chapter 7

Conclusions

The importance of stones in the preservation of cultural heritage

Our heritage partly defines what we are as a society. The state of preservation of our heritage reflects that. Preservation of our natural and cultural heritage is part of sustainable development for future generations. Europe is very rich in cultural heritage, as is the case in Spain and is exemplified by Salamanca, which has a large list of important sites that need to be preserved, many of them built in stone. Restoring a monument or a historic building is inevitably very expensive. Sometimes these sites are private property; other times, they belong to public administrations. However, what was a jewel centuries ago, sometimes, now, is not regarded as being worth kept. There are already several World Heritage Sites that have lost their designations, due to neglect, armed conflicts, natural hazards or destruction of the sites by people. Extreme examples are the destruction of the gigantic statues of the Buddha at Bamiyán (Afghanistan), completely destroyed after the Taliban attack in 2001, and the site of the ancient city of Palmyra (Syria), partially destroyed by the DAESH attacks in 2016 and 2017 (Pereira, 2015; http://unesdoc.unesco.org/images/0011/001140/114044e.pdf#page=134).

In Europe, there is a network made up of more than 400 active organisations in the field of cultural heritage as well as national branches such as Hispania Nostra. This Spanish association has, since 2007, maintained a list of endangered natural and cultural sites, *La Lista Roja del Patrimonio* (the Heritage Red List) (http://listarojapatrimonio.org/). Spain is the third country in the world for the number of protected World Heritage Sites: currently 47. But the country has also more than 15,000 cultural heritage sites that should be looked after. The members of Hispania Nostra have identified more than 500 sites (monuments, historic buildings, vernacular architecture) in Spain that, if no restoration is implemented, will disappear. This list is built based on the historic and architectonic importance of the buildings as well as the state that the buildings are in at present. It also considers the social and economic importance of their possible disappearance, so this list has a strong link with society. However, no mention is made of the importance of the original building materials: generally, the local natural stone.

The links between heritage stone groups and associations around the world can be a step towards appropriate restoration interventions actions on the affected and/or endangered buildings. It is an obligation for researchers and leaders of UNESCO projects, such as the International Geoscience Programme (IGCP), to alert public authorities of the possible consequences of not following a strict protocol in the restoration of historic buildings, particularly for World Heritage Sites, but also more widely.

As far as possible, the same natural stone from the original source should be used to minimise adverse consequences for the historic and architectural heritage. If that is impossible, a very similar material is required. Use of inappropriate stone or treatment with incompatible mortars can have structurally, and very expensive, damaging consequences or be aesthetically unacceptable.

Inappropriate use of stone usually arises because of a lack of information and awareness amongst commissioners of, and those who are responsible for, work. Budget constraints often lead to the selection of cheaper alternatives, but that can be a false economy that leads to much higher future costs. Initial selection of suitable stone is important, but inappropriate attempts at repair exacerbate problems even in some World Heritage Sites. Selected examples from Western Europe illustrate the inappropriate use of mortar and replacement of stone, as has been pointed out in this book.

The Global Heritage Stone initiative was launched to encourage standard reporting of technical data on, and to improve recognition of, the most important heritage stones internationally; to promote their proper use in construction, maintenance and repair; and to stress the need to safeguard important stone resources for future use. An increasingly important consideration today is the possibility of reconstruction of world heritage that has been destroyed or severely damaged due to armed conflicts (Pereira, 2015). In those cases, the original and proper stone should be used to represent the original treasures as closely as possible.

All too often, important places recognised as World Heritage Sites or cities show a deplorable state of at least some of their monuments and historic buildings. This often happens when the historic buildings were constructed of local stone but that stone, for any reason, is no longer available. Frequently, historic quarries that have been inactive for years have become waste dumps, where there is no possibility to study the state of the stone, let alone complete the extraction of that stone even at low quantities for suitable replacement. In these cases, sometimes due to lack of knowledge, lack of interest or cost restraints, the architects and those responsible for restoration decide to use a different stone, if any action takes place at all. In this book, it has been made clear how architects centuries ago were very concerned about the stones they were using in the construction of the historic cities. This is why, after so many centuries, these cities are still regarded as works of art. This book also describes stones that were used in the building of Salamanca to maintain this magnificent city, which fully deserved recognition as a UNESCO World Heritage Site, mainly for its architecture.

A main objective of the present book, besides being a reference source of information on the described stones, is to encourage architects and restoration companies to use the most appropriate materials in the restoration of historic buildings, not only in Salamanca, but in all the cities and towns that have outstanding architecture built in stone. There is an overriding need to respect UNESCO policy regarding the maintenance of a World Heritage Site. Hopefully this book and future volumes on this subject will help to achieve this goal.

Glossary

Accessory minerals   any mineral in small amounts (between 2% and 5%) in an igneous rock not essential to the naming of the rock.

Amphibole   group of minerals appearing in the form of prisms or needle-like crystals.

Anisotropy   physical property of an object or substance according to which qualities such as elasticity, temperature, conductivity, light propagation speed, resistance, etc. have different values when measured in different testing directions.

Anthropic activity   action resulting from or influenced by human activity or intervention on the environment or on building materials.

Apatite   mineral with the following chemical formula: $Ca_5(PO_4)_3$ (F,Cl,OH). Apatite is well crystallised and it has a variable colour, though colourless and greenish crystals are prominent.

Arkose   a type of sandstone containing at least 25% feldspar; it can also be called feldspathic sandstone.

Ashlar   finely dressed stones, generally parallelepiped, used in building, set together in regular patterns and without cementing material between the blocks, unlike masonry.

Baroque style   artistic style (1590–1720) during which authors expressed misfortune and dramatism. Works reflected exaggerated emotion and represented details very realistically. They sometimes emulated a theatre stage set.

Biotite   common phyllosilicate mineral within the mica group, with the approximate chemical formula K(Mg,Fe) 3AlSi 3O 10(F,OH) 2.

Density   dry mass of a rock per unit of actual *in situ* rock volume, including porosity.

Calzo   Spanish term referring to a stone in the joint of two other rocks that limits their movement or which acts as a point of support increasing seat surface.

Charnockite   once considered igneous rocks, today it is proven that most are metamorphic rocks, formed under very high temperature and pressure. The main characteristic is the presence of orthopyroxene. The name derives from "Charnock", considered the founder of Calcutta, India, whose tombstone is made up of this stone. Although many charnokites come from India, similar rocks can be found in Norway, France, Sweden, Germany, Scotland and North and South America.

Chlorite   generic name for silicates of aluminium of the group of phyllosilicates. Iron is usually predominant in the chemical formula and it is sometimes represented by biotite formed from alteration of chlorite.

Clayey matrix   the less-than-4-mm fine-grained materials that surround larger grains in a sedimentary rock.

Cloister   gallery of columns surrounding a garden or an inner courtyard.

Conglomerate   rock consisting of different size particles of one or more substances (sand, clay, rock grains, etc.) usually cemented but in which the most conspicuous grains are of pebble size or larger.

Cordierite   magnesium iron aluminium silicate mineral. Its chemical formula is $(Mg,Fe)_2Al_4Si_5O_{18}$.

Dome   architectural element generally resembling the upper half of a sphere, which was often placed over the central aisle of a church to ventilate and let air or light in.

Durability   behaviour of rocks in weathering processes.

Essential minerals   minerals used to assign a classification name to a rock.

Facies   an association of rocks (usually sedimentary) which reflect the environmental conditions under which they were formed. Facies may change laterally or vertically reflecting change from one environmental set of conditions to another.

Feldspar   a group of aluminosilicate minerals that make up more than half of the Earth's crust and are the main component in granitic rocks and an important constituent in sedimentary arkoses.

Foliation   arrangement of leaf-like layers in a rock occurring when the rock is subjected to extremely high pressure. It is related to the tectonics and metamorphism that affect rocks.

Frieze   wide central section that is part of the entablature of a classical building, which lies horizontally between the cornice and the architrave. It is usually decorated with sculptures and bas-reliefs and sculptures.

Gabarro   in quarrying, an informal Spanish word referring to microgranular enclaves in rocks, mainly in granites.

Gothic style   artistic style in Western Europe lasting from the last centuries of the Middle Age, mid-12th century, until the beginning of Renaissance (15th century in Italy), and even persisting well into the latter part of the 16th century in some places.

Granitoid   granite-like rocks, although their composition is to some extent different mainly in terms of silica content.

Granodiorite   phaneritic-textured igneous plutonic rock (with crystals that can be easily seen) similar to granite. It mainly contains quartz (> 20%) and feldspar but, unlike granite, it contains more plagioclase feldspar than orthoclase feldspar.

Hercynian orogeny   also known as the Variscan orogeny, this is a geological event that took place during the Late Palaeozoic (between 380 and 280 million years ago) during the continental collision between Euramerica and Gondwana, forming into the supercontinent Pangea, and causing extensive mountain formation and igneous (e.g. granite formation) and metamorphic activity near the line of collision.

Inselberg   isolated rocky hill rising abruptly from a flat plain or gently sloping area. It can also be described as a rocky residual relief formed by erosion.

Larvikite   igneous rock, extracted from Larvik (Norway), characterised by the presence of large crystals of feldspar with a peculiar iridescence when polished, giving the rock commercial names such as "Blue Pearl".

Lithology   the composition and textures of various types of rocks.

Macroscopic   scale on which objects or phenomena are large enough to be visible with the naked eye.

Marble   sedimentary rock consisting chiefly of calcium carbonate (limestone), which has undergone metamorphic recrystallisation. The term is often used incorrectly commercially for any type of rock which can be polished to a good reflective finish.

Masonry   the building of structures from individual units of different sizes and shapes that are bound together by mortar or hydraulic lime (in contrast to ashlar).

Mesoscopic scale large enough to be examined with the naked eye, but not big enough to examine the whole body as a single entity.

Mica group of minerals consisting of very thin, shiny, soft and flexible layers.

Monzodiorite a coarse-grained igneous rock made of plagioclase feldspar as the main mineral, and orthoclase feldspar, hornblende and biotite in minor proportions.

Muscovite mineral of the silicate group, phyllosilicate subgroup. Muscovite consists of aluminium with varying proportions of potassium, magnesium, chrome and other chemical elements.

Opal a mineral consisting of amorphous silica containing water molecules and without crystal structure.

Open pit quarry generally surface mining area from which industrial stones are quarried (ornamental stones or aggregates).

Outcrop exposed bedrock, not covered by soil or other rocks; any place where bedrock is visible on the surface of the Earth and its external features of colour and structure can be seen.

Petrography branch of science that studies, describes and classifies rocks.

Phaneritic texture of a rock in which the principal constituents are crystals visible to the naked eye.

Weathering the physical breakup (disintegration) of rocks and soils through large temperature oscillations or abrasion by materials carried in water, ice or wind, during which rocks are broken up progressively into small grains of sand.

Plagioclase a series of silicate minerals, a very common form of feldspar in many rocks including granites.

Plateresque style Spanish architectural style which was only expressed during the Spanish Renaissance. According to historic documents, it started at the beginning of the 15th century and lasted for two more centuries.

Plutonic rocks intrusive igneous rock that solidified from a melted substance at great depth.

Porphyritic texture of igneous rocks containing large crystals embedded in a fine groundmass of minerals.

Puddingstone conglomerate that consists of rounded pebbles of different sizes with siliceous or carbonated cement.

Quartz a mineral consisting of silicon dioxide occurring in colourless and transparent or coloured hexagonal crystals or in crystalline masses.

Rapakivi granite   granite with orbicular texture made up of large crystals of feldspar surrounded by plagioclase and amphibole. The origin of this granite is in Finland and the meaning of the name is "rotten" or "crumbly" rock, despite the fact that much of it is a very good construction material.

Resistance   measure of how well rocks resist processes such as compression, tension and flexure.

Romanesque style   architecture style of mediaeval age in Europe, with the characteristics of circular constructions with semi-circular arches. The style is dated to around the 11th century, passing to Gothic style in the 12th century.

Secondary minerals   minerals that form later than an original igneous rock through the alteration of a pre-existing mineral.

Sericite   fine-grained clayey minerals that often forms aggregates in rocks.

Strength   equivalent to resistance.

Tourmaline   glassy mineral of variable colour consisting of aluminium silicate with sodium, boron, magnesium and other chemical components, which can occur as accessory minerals in igneous and metamorphic rocks.

Zircon   mineral of variable colour consisting of zirconium silicate, the chemical formula of which is $ZrSiO_4$.

References

Agostini, A., Barello, F., Borghi, A. & Compagnoni, R. (2017) The white marble of the Arch of Augustus (Susa, North-Western Italy): Mineralogical and petrographic analysis for the definition of its origin. *Archaeometry*, 59(3), 395–416.

Angotzi, G., Bramanti, L., Tavarini, D., Gragnani, M., Cassiodoro, L., Moriconi, L., Saccardi, P., Pinto, I., Stacchini, N. & Bovenzi, M. (2015) World at work: Marble quarrying in Tuscany. *Occupational and Environmental Medicine*, 62, 417–421.

ASTM (2004) D4404–84, Standard Test Method for Determination of Pore Volume and Pore Volume Distribution of Soil and Rock by Mercury Intrusion Porosimetry. *ASTM International, 100 Barr Harbor Drive, West Conshohocken, PA 19428, USA.*

ASTM (2013) D5312, Standard Test Method for Evaluation of Durability of Rock for Erosion Control Under Freezing and Thawing Conditions. *Copyright ASTM International, 100 Barr Harbor Drive, West Conshohocken, PA 19428, USA.*

ASTM (2015a) C-1721–15, Standard Guide for Petrographic Examination of Dimension Stone. *Copyright ASTM International, 100 Barr Harbor Drive, West Conshohocken, PA 19428, USA.*

ASTM (2015b) C-880 / C880M-15, Standard Test Method for Flexural Strength of Dimension Stone. *ASTM International, West Conshohocken, PA, 19428, USA.*

ASTM (2015c) C-616, Standard Specification for Sandstone Dimension Stone. *Copyright ASTM International, 100 Barr Harbor Drive, West Conshohocken, PA 19428, USA.*

ASTM (2015d) C-615, "Standard Specification for Granite Dimension Stone". *Copyright ASTM International, 100 Barr Harbor Drive, West Conshohocken, PA 19428, USA.*

ASTM (2017) C170 / C170M-17, Standard Test Method for Compressive Strength of Dimension Stone. *ASTM International, West Conshohocken, PA, 19428, USA.*

ASTM (2018) C97 / C97M-18, Standard Test Methods for Absorption and Bulk Specific Gravity of Dimension Stone. *ASTM International, West Conshohocken, PA.*

Azofra, E. & Pérez-Hernández, M. (2013) *Loci et imagines/Images and Places.* Edited by University of Salamanca, Salamanca, Spain.

Blanco, J.A. (1991) Weathering process on Tertiary basins [in Spanish]. In: Blanco, J.A., Molina, E. & Martín Serrano, A. (eds.) *Weathering and Paleoweathering on the West Peninsular Morphology: Hercynic Basins and Tertiary Basins.* Monografías Sociedad Española de Geomorfología, 6, Geoforma ediciones, Logroño, Spain, pp. 45–67.

Cooper, B., Marker, B., Pereira, D. & Schouenborg, B. (2013) Establishment of the "Heritage Stone Task Group" (HSTG). *Episodes,* 36(1), 8–10.

Díez Balda, M.A. (1986) *El Complejo Esquisto Grauváquico, las Series Paleozoicas y la Estructura Hercínica al Sur de Salamanca.* Ediciones Universidad de Salamanca, Salamanca, Spain, p. 162.

Díez Balda, M.A., Vegas, R. & González Lodeiro, F. (1990) Structure, Part IV: Central Iberian Zone. In: *Pre-Mesozoic Geology of Iberia.* Springer Verlag, Berlín Heidelberg, Germany, pp. 172–188.

Douet, J. (2015) *European Quarry Landscapes.* ISBN: 978-84-88220-26-4. Available from: http://media.globalheritagestone.com/2016/11/LIFE-Teruel-publication-7-13.pdf

EN (2008) Natural Stone: Denomination Criteria. *EN 12440:2008 Comite Europeen de Normalisation.* Management Centre: rue de Stassart, 36 B-1050, Brussels, Belgium.

García de los Ríos Cobo, J.I. & Báez Mezquita, J.M. (2001) La Piedra en Castilla y León. *Junta de Castilla y León.* p. 345 [in Spanish].

Garcia-Talegón, J., Iñigo, A.C., Alonso-Gavilán, G. & Vicente-Tavera, S. (2015) *Villamayor Stone (Golden Stone) as a Global Heritage Stone Resource from Salamanca.* Geological Society, London, UK. Special Publications, 407. pp. 109–120.

Gentili, R., Sgorbati, S. & Baroni, C. (2011) Plant species patterns and restoration perspectives in the highly disturbed environment of the Carrara marble quarries (Apuan Alps, Italy). *Restoration Ecology,* 19, 32–42.

González Neila, C., Pereira, D. & Baltuille, J.M. (2017) Granitos de Salamanca. Un ejemplo perfecto para el reconocimiento de la piedra natural como recurso patrimonial y su uso en restauración. *Boletín Geológico y Minero,* 128(2), 677–688.

Harrell, J.A. & Storemyr, P. (2009) Ancient Egyptian quarries: An illustrated overview. In: Abu-Jaber, N., Bloxam, E.G., Degryse, P. & Heldal, T. (eds.) *QuarryScapes: Ancient Stone Quarry Landscapes in the Eastern Mediterranean.* Geological Survey of Norway. Special Publication, 12, Norway, pp. 7–50.

Hevia, D. (1860) *Mapa geográfico-estadístico-itinerario de la provincia de Salamanca.* Litografía Marquerie, Madrid, Spain.

ICOMOS-ISCS (2008) Internacional scientific comitee for stone: Illustrated glossary on stone deterioration patterns. p. 78. Ateliers 30 Impression, Champigny/Marne, France.

Karagiannis, N., Karoglou, M., Balokas, A. & Moropoulou, A. (2016) Building materials capillary rise coefficient: Concepts, determination and parameters involved. In: Delgado, J.M.P.Q. (ed.) *New Approaches to Building Pathology and Durability*. Springer, Singapore, pp. 27–44.

Karoglou, M., Moropoulou, A., Giakoumaki, A. & Krokida, M.K. (2005) Capillary rise kinetics of some building materials. *Journal of Colloid and Interface Science*, 284(1), 260–264.

Kramar, S., Bedjanič, M., Mirtič, B., Mladenović, A., Rožič, B., Skaberne, D., Gutman, M., Zupančič, N. & Cooper, B. (2014) Podpeč limestone: A heritage stone from Slovenia. In: Pereira, D., *et al.* (eds.) *Towards International Recognition of Building and Ornamental Stones*. Geological Society, London, UK. Special Publications, 407. pp. 219–231.

Le Maitre, R.W. (2002) *Igneous Rocks: A Classification and Glossary of Terms*. Cambridge University Press, Cambridge, MA.

López Plaza, M., García de los Ríos Cobo, J.I., López Moro, F.J., González Sánchez, M., Íñigo, A.C., Vicente Tavera, S. & Jiménez Fuentes, E. (2009) La utilización del granito de Los Santos en la ciudad de Salamanca. *Studia Geologica Salmanticensia*, 45(1), 7–40.

López Plaza, M., González Sánchez, M., García de los Ríos Cobo, J.I., Cortázar Estíbaliz, J., Íñigo, A.C., Vicente Tavera, S. & López Moro, F.J. (2007a) La utilización de rocas vaugneríticas en los monumentos de Salamanca. *Studia Geologica Salmanticensia*, 43(1), 115–142.

López Plaza, M., Sánchez, M.G. & Iñigo, A.C. (2007b) La utilización del leucogranito turmalinífero de Martinamor en los monumentos de Salamanca y Alba de Tormes. *Studia Geologica Salmanticensia*, 43(2), 247–280.

Los Pueblos Más Bonitos de España (2018) Available from: https://www.lospueblosmasbonitosdeespana.org/

Macarro, C. & Alario, C. (2012) *Los Orígenes de Salamanca. El poblado protohistórico del Cerro de San Vicente*. Centro de estudios Salmantinos, Salamanca, Spain.

Madoz, P. & Sagasti, L. (1845) *Diccionario geográfico-estadístico-histórico de España y sus posesiones de ultramar*. Est. Literario-Tipográfico de P. Madoz y L. Sagasti. p. 11.668.

Marini, P. & Mosseti, C. (2006) Natural stones used in a Royal House of Piedmont (Italy). In: Heritage, Weathering and Conservation, Fort et al. (eds.), CRC Press, Taylor and Francis Group. pp. 895–900.

Marker, B. (2015) Procedures and criteria for the definition of Global Heritage Stone Resources. In: Pereira, D., *et al.* (eds.) *Towards International Recognition of Building and Ornamental Stones*. Geological Society, London, UK. Special Publications, 407. pp. 5–10.

Menéndez Bueyes, L.R. (2000–2001) El puente romano de Salamanca y su contexto historico (A proposito de CIL II 4685). *Memorias de Historia Antigua*, 21–22, 149–183.

Molina, E., García Talegón, J. & Alonso Gavilán, G. (2009) Papel de la porosidad en el proceso de silicificación del borde occidental de la Cuenca Terciaria del Duero. *Revista de la Sociedad Geológica de España*, 22(3–4), 145–154.

Neiva, A.M.R., Silva, M.M.V.G. & Gomes, M.H. (2007) Crystal chemistry of tourmaline from Variscan granites, associated tin-tungsten- and gold deposits, and associated metamorphic and metasomatic rocks from northern Portugal. *Neues Jahrbuch für Mineralogie – Abhandlungen*, 184(1), 45–76.

Nespereira, J., Blanco, J.A., Yenes, M. & Pereira, D. (2010) Opal cementation in tertiary sandstones used as ornamental stones. *Engineering Geology*, 115, 167–174.

Ordaz, J. (1983) Características físicas y alterabilidad de la piedra de Villamayor (Salamanca). *Materiales de Construcción*, 190–191, 85–96.

Pereira, D. (2015) The value of global heritage stone resource designation. *International Congress Les inventaires du géopatrimoine Procceedings*. Toulouse, France, pp. 279–285.

Pereira, D. & Cooper, B. (2013) Building stone as a part of a World Heritage site: "Piedra Pajarilla" Granite and the city of Salamanca. In: Cassar, J., et al. (eds.) *Stone in Historic Buildings: Characterization and Performance*. Geological Society. Special Publications, London, UK, 391, 7–16.

Pereira, D. & Cooper, B. (2015) A Global Heritage Stone Province in association with the UNESCO World Heritage City of Salamanca, Spain. In: Lollino, G., et al. (eds.) *Engineering Geology for Society and Territory*. Volume 5. Springer International Publishing, Switzerland. pp. 205–208.

Pereira, D., Gimeno, A. & del Barrio, S. (2015) Piedra Pajarilla: A candidacy as a Global Heritage Stone Resource for Martinamor granite. In: Pereira, D., et al. (eds.) *Towards International Recognition of Building and Ornamental Stones*. Geological Society, Special Publication, London, UK, 407, 93–100.

Pereira, D., González Neila, C. & del Arco, A. (2018) *Salamanca, ciudad de piedra y patrimonio/Salamanca, city of stone and heritage*. Diputación de Salamanca, Spain, p. 80.

Pereira, D. & Marker, B. (2016a) The value of original natural stone in the context of architectural heritage. *Geosciences*, 6(1). doi:10.3390/geosciences6010013

Pereira, D. & Marker, B. (2016b) Repair and maintenance of natural stone in historical structures: The potential role of the IUGS Global Heritage Stone initiative. *Geoscience Canada*, 43, 5–12.

Pereira, D. & Pérez Castro, P. (2017) Art Museums: A good context for outreach activities on natural stones and heritage. *Geoheritage*. doi:10.1007/s12371-017-0265-9

Pereira, M.D. & Rodríguez-Alonso, M.D. (2000) Duality of cordierite granites related to melt-restite segregation in the Peña Negra Anatectic Complex, central Spain. *Canadian Mineralogist*, 38, 1329–1346.

Pérez Delgado, T. (2002) *Los Arapiles. La batalla y su entorno*. Diputación de Salamanca, Salamanca, Spain.

Portal-Monge, Y. (1988) *La torre de las campanas de la catedral de Salamanca*. Universidad de Salamanca, Salamanca, Spain, p. 198.

Rodríguez de Ceballos, A. (1978) La torre de la catedral nueva de Salamanca. *Boletín del Seminario de Estudios de Arte y Arqueología*, 44, 245–256.

Stock, D. (2012) Old city of Salamanca. *Discovering Stone*, 21, 54–56.

Touret, J.L.R. & Bulakh, A.G. (2016) The Russian contribution to the edification of the Napoleon tombstone in Paris. *Вестник СПбГУ*. Сер. 15. 2016, 3, 70–83. doi:10.21638/11701/spbu15.2016.306

UNESCO (2016) *Basic Texts of the 2003 Convention for the Safeguarding of the Intangible Cultural Heritage*. UNESCO, Paris, France. p. 132.

Vielva, C. (2001) *La arenisca de Villamayor en recubrimientos de fachada*. Doctoral Thesis. Madrid, Spain.

Wikström, A., Pereira, D., Lundqvist, T. & Cooper, B. (2015) The Dala (Älvdalen) porphyries from Sweden. *Episodes*, 38(2), 114–117.